Critical Issues in Mathematics Education

Philip Clarkson · Norma Presmeg
Editors

Critical Issues in Mathematics Education

Major Contributions of Alan Bishop

Springer

Editors
Philip Clarkson
Faculty of Education
Australian Catholic University
Fitzroy VIC 3065
Australia
p.clarkson@patrick.acu.edu.au

Norma Presmeg
Illinois State University
Department of Mathematics
313 Stevenson Hall
Normal IL 61790-4520
USA
npresmeg@msn.com

ISBN: 978-0-387-09672-8 e-ISBN: 978-0-387-09673-5

Library of Congress Control Number: 2008930313

Printed on acid-free paper

springer.com

Contents

Section I Introduction

1 **Developing a Festschrift with a Difference** 3
Philip Clarkson and Norma Presmeg

2 **In Conversation with Alan Bishop** 13
Philip Clarkson

Section II Teacher Decision Making

3 **Decision-Making, the Intervening Variable** 29
Alan J. Bishop

4 **Teachers' Decision Making: from Alan J. Bishop to Today** 37
Hilda Borko, Sarah A. Roberts and Richard Shavelson

Section III Spatial Abilities, Visualization, and Geometry

5 **Spatial Abilities and Mathematics Education – A Review** 71
Alan J. Bishop

6 **Spatial Abilities Research as a Foundation for Visualization in Teaching and Learning Mathematics** 83
Norma Presmeg

7 **Spatial Abilities, Mathematics, Culture, and the Papua New Guinea Experience** 97
M.A. (Ken) Clements

Section IV Cultural and Social Aspects

8 **Visualising and Mathematics in a Pre-Technological Culture** 109
Alan J. Bishop

9 **Cultural and Social Aspects of Mathematics Education: Responding to Bishop's Challenge** 121
Bill Barton

10 **Chinese Culture, Islamic Culture, and Mathematics Education** 135
Frederick Leung

Section V Social and Political Aspects

11 **Mathematical Power to the People** 151
Alan J. Bishop

12 **Mathematical Power as Political Power – The Politics of Mathematics Education** 167
Christine Keitel and Renuka Vithal

Section VI Teachers and Research

13 **Research, Effectiveness, and the Practitioners' World** 191
Alan J. Bishop

14 **Practicing Research and Researching Practice** 205
Jeremy Kilpatrick

15 **Reflexivity, Effectiveness, and the Interaction of Researcher and Practitioner Worlds** 213
Kenneth Ruthven

Section VII Values

16 **Mathematics Teaching and Values Education – An Intersection in Need of Research** 231
Alan J. Bishop

17 **Valuing Values in Mathematics Education** 239
Wee Tiong Seah

Index 253

Contributors

Bill Barton
Dept. of Mathematics, The University of Auckland, Private Bag 92019, Auckland Mail Centre, Auckland 1142, New Zealand, b.barton@auckland.ac.nz

Alan J. Bishop
Faculty of Education, Monash University, Wellington Road, Clayton, Victoria 3168, Australia, alan.biship@education.monash.edu.au

Hilda Borko
School of Education, Stanford University, 485 Lasuen Mall, Stanford, CA 94305-3096, USA, hildab@suse.stanford.edu

Philip C. Clarkson
Faculty of Education, Australian Catholic University, Fitzroy VIC 3065, Australia, p.clarkson@patrick.acu.edu.au

M. A. (Ken) Clements
Department of Mathematics, Illinois State University, Normal, IL 61790-4520, United States of America, clements@ilstu.edu

Christine Keitel
Fachbereich Erziehungswissenschaft und Psychologie, Freie Universität Berlin, Habelschwerdter Allee 45, 14195 Berlin, Germany, keitel@zedat.fu-berlin.de

Jeremy Kilpatrick
105 Aderhold Hall, University of Georgia, Athens, GA 30602-7124, United States of America, jkilpat@uga.edu

Frederick Leung
Chair Faculty of Education, The University of Hong Kong, Pokfulam Road, Hong Kong, hraslks@hkucc.hku.hk

Norma Presmeg
Illinois State University, Department of Mathematics, 313 Stevenson Hall, Normal IL 61790-4520, USA, npresmeg@msn.com

Sarah A. Roberts
School of Education, 249 UCB University of Colorado, Boulder, CO 80309-0249, USA, Sarah.A.Roberts@Colorado.edu

Kenneth Ruthven
Faculty of Education, University of Cambridge, 184 Hills Road, Cambridge CB2 8PQ, United Kingdom, kr18@hermes.cam.ac.uk

Wee Tiong Seah
Faculty of Education, Monash University (Peninsula Campus), PO Box 527, Frankston, Vic. 3199, Australia, WeeTiong.Seah@Education.monash.edu.au

Richard Shavelson
School of Education, 485 Lasuen Mall, Stanford University, CA 94305-3096, USA, richs@stanford.edu

Renuka Vithal
Dean of Education, Faculty of Education, University of KwaZulu-Natal, Edgewood Campus, Private Bag X03, Ashwood 3605, South Africa, vithalr@ukzn.ac.za

Section I
Introduction

Chapter 1
Developing a Festschrift with a Difference

Philip Clarkson and Norma Presmeg

A Festschrift is normally understood to be a volume prepared to honour a respected academic, reflecting on his or her significant additions to the field of knowledge to which they have devoted their energies. It is normal for such a volume to be composed of contributions from those who have worked closely with the academic, including doctoral students, and others whose work is also known to have made important contributions within the same areas of research.

It was the dearth of volumes of this type in the area of mathematics education research that Philip Clarkson and Michel Lokhorst, then a commissioning editor with Kluwer Academic Publishers, started to discuss some 5 years ago. This discussion point was embedded in a broader conversation that lamented the fact that little was published that kept a trace of how ideas developed over time in education, and in mathematics education in particular. Associated with this notion was how we as a community were not very good at linking the development of ideas with the people who had worked on them, and the individual contexts within which their thinking occurred. We wondered whether something should be done to draw attention to this issue. One way to do that was to begin the task of composing a Festschrift, but with a difference.

In thinking through the implications of this proposition, it seemed useful to structure the volume in such a way that perhaps more could be achieved than by just initiating a call for contributions to honour a colleague who had made a long and important contribution to mathematics education. We wondered whether a structure could be developed for the proposed volume that emphasised the following:

- the ideas of the honoured academic that she or he had developed,
- where and how they were developed, and
- what became of those ideas once they were published and taken up, or not taken up, by the community of scholars that were working in that particular area, in this case mathematics education.

P. Clarkson
Faculty of Education, Australian Catholic University, Fitzroy VIC 3065, Australia
e-mail: p.clarkson@patrick.acu.edu.au

P. Clarkson, N. Presmeg (eds.), *Critical Issues in Mathematics Education*,

We decided that indeed such a project should be initiated. It was relatively easy to decide to focus on Alan Bishop's contributions to mathematics education over the last 40 years, which are still continuing. This, then is the goal of this volume.

The purpose of this volume is twofold, each part of equal weight, although the second component has given the impetus and structure for the volume. The first is to put into perspective the contribution that (now Emeritus) Professor Alan Bishop has made to mathematics education research beginning in the 1960s. The other is to review six critical issues that have been important in the establishment of mathematics education research over the last 50 years, including updating to some extent current developments in each of these areas. The volume was planned to make a valuable contribution to the ongoing reflection of mathematic education researchers world wide, but also to address topics relevant to policy makers and teacher educators who wish to understand some of the key issues with which mathematics education has been and still is concerned. However all ideas develop within an historical context. Hence in various places within this volume comment is made with regard to the contexts within which Bishop's contributions to these research issues were made.

Bishop's contributions can be conveniently outlined through a consideration of the following six issues as they relate to mathematics education research:

- Teacher decision-making
- Spatial abilities, visualization and geometry
- Cultural and social aspects of mathematics education
- Socio political issues for mathematics education
- Teachers and research
- Values and teaching mathematics.

The structure of the volume has been developed around these six issues, each issue being the focus of a section of the volume. Each section has three or four components. The first component of each section is a brief introduction that positions and gives a context for the Bishop article reprinted in the section.

The second component of each section is a reprint of a particular "key" journal article or book chapter that Bishop published. Each key article has been chosen to typify his contribution to the ongoing research on that issue. These articles were selected in conversation with Bishop.

The final component of each section consists of one or two invited chapters from selected authors. We chose authors who had either worked directly with Bishop, or had worked with the ideas canvassed in their section.

Authors were asked to use the Bishop key article for their section as a focus for a commentary on that issue in mathematics education. We anticipated that the authors would use the key articles in different ways: perhaps as a starting point to develop a dialog with the article in some way, or to take the key article and map out how the ideas have or have not been taken up in succeeding years, or to look back to what preceded the publication of the article and place it in an historical context, or to start in a completely different place and come back to the notions discussed in the key article. The aim was for the ideas embedded in the key Bishop article to be central in the formation of each contributed chapter. We hoped that a number of approaches would be used which would give the volume a feel of variety and surprise, bound together by the brief introductory components of each section. We believe this has been achieved.

When colleagues who have worked directly with Bishop in some way, or have worked with his ideas, are asked to contribute to such a volume as this, there is a danger of the volume becoming just a set of personal reflections about him. At times documenting publicly the appreciation of and esteem in which we hold colleagues is most appropriate, and perhaps not done often enough. But more than this was envisaged for this volume. We were also aiming for a scholarly contribution to the literature. We thought that this was the best way we could honour Bishop's legacy. Hence we wanted to do both; record a little of the community's personal appreciation of Bishop's contributions over many years, but also try to make some scholarly advances in our thinking.

We had originally envisaged having two separate authors contributing to each section, at first working independently and then commenting on each other's chapters. We thought that in this way we would have to some extent a divergent yet focussed commentary on each issue, and indeed Bishop's contributions to mathematics education research. However, as can be seen, this did not always prove feasible. At times we took up the suggestion from particular colleagues that they develop a joint chapter. We would also note with appreciation that although our friend and colleague (and one of Bishop's doctoral students) Chien Chin had agreed to contribute a chapter for the last section, illness in the end prevented him from doing so.

We also suggested to the authors that inclusion in their chapter of pertinent anecdotal and/or biographical comments on Alan and his contribution to mathematics education research would not be out of place. This has been done in different ways by different authors, and enlarges the understanding of the contexts in which Bishop worked through his own ideas. As noted in the introduction component of the section dealing with teacher decision-making, Bishop firmly believed that research in education is not a disembodied objective process. Rather the researcher is intimately contained within the research process in various ways, whether those ways are immediately clear to the researcher and others involved in a particular project or not. Hence knowing more about Bishop allows us to know more and understand in different ways his contributions to the research of mathematics education.

Just as it is important to know something of the contexts within which Bishop worked while contributing to these different issues in the ongoing research of mathematics education, it is also useful to know something of the authors who were kind enough to contribute to this volume. The following paragraphs give brief introductions to each author who has contributed to the volume either individually or as part of a team. We also need to acknowledge the help, support and guidance of our editors at Springer, Marie Sheldon and Kristina Wiggins-Coppola, who have worked with us from a very early stage in the process of publishing this volume, and without whose support and insights we would not have made it through the publishing process.

Contributing Authors

Bill Barton

Bill Barton is Head of Mathematics at The University of Auckland, having come to university after a secondary teaching career including bilingual Maori/English mathematics teaching. His research areas include ethnomathematics and mathematics and language. Bill has known Alan since the early 1990s, and regards him as being one of the key influences on his mathematics education research.

Hilda Borko

Hilda Borko is Professor of Education, School of Education, Stanford University. Dr. Borko's research examines the process of learning to teach, with an emphasis on changes in novice and experienced teachers' knowledge and beliefs about teaching and learning, and their classroom practices as they participate in teacher education and professional development programs. Currently, her research team is studying the impact of a professional development program for middle school mathematics teachers which they designed, on teachers' professional community and their knowledge, beliefs, and instructional practices. Many of Alan Bishop's ideas about teacher decision making and the use of video as a tool for teacher learning are reflected in that work.

Philip Clarkson

Philip is Professor of Education at Australian Catholic University where he has taught since 1985. This followed nearly 5 years as Director of a Mathematics Education Research Centre at the Papua New Guinea University of Technology, and prior to that as a lecturer at Monash University and tertiary colleges in Melbourne. He began his professional life as a teacher of mathematics, chemistry, environmental science and physical education in secondary schools. At present he is the Deputy Director of the Mathematics and Literacy Education Research Flagship at Australian Catholic University, the State Coordinator of graduate research programs, teaching general education and mathematics education units in these programs, and tutoring in first year mathematics. He has served as President, Secretary and Vice President (Publications) of the Mathematics Education Research Group of Australasia (MERGA), and was the foundation editor of the association's research journal *Mathematics Education Research Journal.* Major funded research projects in the last 10 years have been: "A longitudinal evaluation of the teacher education programs in Papua New Guinea"; "An evaluation of the computer Navigator Schools Project"; "The impact of language on mathematics learning, particularly for bilingual students"; and "Globalisation and the professional development of mathematics educators". Philip met Alan in 1977, and from then on our paths have regularly crossed. They have had a mutual interest in education in Papua New Guinea, a particular context for discussions of language and cultural issues. They have also worked together in various ways, particularly on the project "Values and Mathematics" and in the running of the 1995 ICME Regional Conference.

M. A. (Ken) Clements

Since 2005, M. A. (Ken) Clements has been a Professor within the Mathematics Department at Illinois State University. He was in charge of mathematics education at Monash University between 1974 and 1982, and subsequently held positions in mathematics education at Deakin University (1987–1993), the University of Newcastle (NSW) (1993–1997), and Universiti Brunei Darussalam (1997–2004). Ken has authored and/or edited many articles, chapters, and books on mathematics education. Ken first met Alan Bishop when Alan came to Monash University as a visiting Research Fellow during the second half of 1977. He has subsequently worked with Alan on many projects, including as co-editor of two international handbooks on mathematics education.

Christine Keitel

Christine Keitel is Professor for Mathematics Education at Free University of Berlin. At present she is serving a second term as Vice-President (Deputy Vice Chancellor) of the university, responsible for restructuring of teaching and research. Her major research areas are comparative studies on the history and current state of mathematics education in various European and Non-European countries, on social practices of mathematics, on values of teachers and students, on "mathematics for all" and "mathematical literacy", on equity and social justice, on learners' perspectives on classroom practice, and on internationalization and globalization of mathematics education.

She was a member of the International Group BACOMET (Basic Component of Mathematics Education for Teachers) 1985–2005 and its director 1989–1993, director of the NATO- Research Workshop on "Mathematics Education and Technology" 1993–1994, a member of the Steering Committee of the OECD-project "Future Perspectives of Science, Mathematics and Technology Education" (1989–1995), Expert Consultant for the Middle-School Reform Project in PR China in 1990, for the Indonesian Ministry of Education in 1992, and for the TIMSS-Video-Project and Curriculum-Analysis-Project (1993–1995). She is member of editorial boards of several journals for curriculum and mathematics education and on the Advisory Board of Kluwer's Mathematics Education Library. Together with David Clarke and Yoshinori Shimizu she started the international LPS-project "Mathematics Classroom Practice: The Learners' Perspective" in

1999, which represents a collaboration of academics of 15 countries around the world (www.edfac.unimelb.edu.au/DSME/lps/). She is leader of the German team of LPS.

She was a founding member, National Coordinator, and Convenor/President of IOWME (International Organisation of Women and Mathematics Education) 1988–1996; Vice-president, Newsletter Editor and President of CIEAEM (Commission Internationale pour l'Etude et l'Amélioration de l'Enseignement des Mathématiques) 1992–2004; and member of the International committee of PME (International Group for Psychology and Mathematics Education) 1988–1992. As a guest professor she has lectured and researched at research institutions and universities around the world, in particular in Southern Europe, USA, Australia and South Africa. In 1999 she received an Honorary Doctorate of the University of Southampton, UK and the Alexander-von-Humboldt/South-African-Scholarship Award for undertaking capacity building in research in South Africa.

Jeremy Kilpatrick

Jeremy Kilpatrick is Regents Professor of Mathematics Education at the University of Georgia. He holds an honorary doctorate from the University of Gothenburg, is a National Associate of the National Academy of Sciences, received a 2003 Lifetime Achievement Award from the National Council of Teachers of Mathematics (NCTM), and received the 2007 Felix Klein medal from the International Commission on Mathematical Instruction. His research interests include teachers' proficiency in teaching mathematics, mathematics curriculum change and its history, assessment, and the history of research in mathematics education. He and his family have known Alan Bishop and his family for more than a third of a century, and when their boys were young, each family spent some months in or near the other family's hometown. A treasured memory is of the four boys and four parents walking the South Downs near Eastbourne during the summer of 1976.

Frederick Koon-Shing Leung

Born and raised in Hong Kong, Frederick Leung is Professor of Mathematics Education in the Faculty of Education, at the University of Hong Kong. Frederick obtained his B.Sc., Cert.Ed. and M.Ed. from the University of Hong Kong, and subsequently his Ph.D. from the University of London Institute of Education. Alan Bishop was an external examiner for his Ph.D. thesis. Frederick's major research interests are in the comparison of mathematics education in different countries, and in the influence of different cultures on teaching and learning. He is the principal investigator of a number of major research projects, including the Hong Kong component of the Trends in International Mathematics and Science Study (TIMSS), the TIMSS 1999 Video Study, and the Learner's Perspective Study (LPS).

Norma Presmeg

Jeremy Kilpatrick introduced me to Alan Bishop at a conference on the recently translated publications of Krutetskii, in Athens, Georgia, USA, in 1980. At the time I was completing a Master of Education thesis on Albert Einstein's creativity, in the Department of Educational Psychology at the University of Natal in South Africa. The heart and soul of Einstein's creative thought, by his own admission, lay in his proclivity for visualization. I had been teaching high school mathematics for 12 years, and noticed that there were students of high spatial abilities who were not succeeding in mathematics in their final year of school. All three of the boys singled out wished to pursue careers that involved visualization, namely, architecture, structural engineering, and technical drawing. The current state of their mathematics achievements would not permit these aspirations to be realized. A research goal was born, namely, *To understand more about the circumstances that affect the visual pupil's operating in his or her preferred mode, and how the teacher facilitates this, or otherwise.* Alan Bishop encouraged me to undertake this research on the strengths and pitfalls of visualization in the teaching and learning of mathematics. My 3 years at Cambridge University (1982–1985) pursuing doctoral research under the able and caring supervision of Alan Bishop remain a highlight of my life. The results of this research on visualization in mathematics education were exciting and fascinating. But the association with Alan opened up another significant field. In 1985, Alan was working on the first three chapters of his book, *Mathematical enculturation*, and it was my privilege to serve as a sounding board for his ideas while I waited to defend my dissertation during those summer months. When I returned to South Africa and worked at the University of Durban-Westville for five years (before immigrating to the USA in 1990), the role of culture in mathematics education became a central topic of my concern. Alan Bishop's influence in my professional career has been a significant one. After 10 years at The Florida State University, I moved to Illinois, where I am currently a Professor in the Mathematics Department at Illinois State University.

Sarah Roberts

Sarah Roberts is a doctoral candidate in mathematics curriculum and instruction at the University of Colorado at Boulder. Her research interests include pre-service and inservice teacher education, equity in mathematics, and issues related to English language learners.

Kenneth Ruthven

After working as a secondary-school teacher in Scotland and England, and completing doctoral research at Stirling University, Kenneth Ruthven was appointed to Cambridge University where he worked closely with Alan Bishop for nearly 10 years. It was during this period that Ken joined the Editorial Board of Educational Studies in Mathematics of which Alan was then Editor-in-Chief; some years later Ken was to take on that senior role; and both currently continue to serve as Advisory Editors. Now Professor of Education at Cambridge, Ken's research focuses on issues of curriculum and pedagogy, especially in mathematics, and particularly in the light of social and technological change and of deepening conceptualisation of educational processes. Recent projects have examined technology integration in secondary subject teaching; future commitments include a major project on principled improvement in STEM education.

Wee Tiong Seah

Wee Tiong Seah is a Lecturer in the Faculty of Education, Monash University, Australia. Amongst his several research interests, he is particularly passionate about researching and facilitating effective (mathematics) teaching/learning through promoting teacher/student's socio-cognitive growth (e.g. values) and through harnessing their intercultural competencies. Wee Tiong completed his doctoral research study in 2004 under the supervision of Alan Bishop and Barbara Clarke. If Alan's migration to Australia from Britain in the early 1990s had been a motivation for Wee Tiong to migrate from Singapore in the late 1990s, then he has also been instrumental in socialising Wee Tiong to the mathematics education research community. Over the years, Alan has also become colleague, mentor and friend to Wee Tiong.

Richard J. Shavelson

Richard J. Shavelson is the Margaret Jacks Professor of Education, Professor of Psychology (courtesy), and former I. James Quillen Dean of the School of Education at Stanford University and Senior Fellow in the Woods Institute for the Environment at Stanford. He served as president of the American Educational Research Association; is a fellow of the American Association for the Advancement of Science, the American Psychological Association, the American Psychological Society, and a Humboldt Fellow. His early research focused on teachers' decision making at the same time Alan Bishop was working in this area. His current work includes the assessment of science achievement and the study of inquiry-based science teaching and its impact on students' knowledge structures and performance. Other current work includes: studies of the causal impact of computer cognitive training on working memory, fluid intelligence and science achievement; assessment of undergraduates' learning with the Collegiate Learning Assessment; accountability in higher education; the scientific basis of education research; and new standards for measuring students' science achievement in the National Assessment of Educational Progress (the nation's "report card"). His publications include Statistical

Reasoning for the Behavioral Sciences, Generalizability Theory: A Primer (with Noreen Webb), and Scientific Research in Education (edited with Lisa Towne). He is currently working on a book titled, The Quest to Assess Learning and Hold Higher Education Accountable.

Renuka Vithal

Renuka Vithal is Professor of Mathematics Education and Dean of the Faculty of Education at the University of KwaZulu-Natal, South Africa. She has taught at all levels from preservice to inservice mathematics teacher education programs, in postgraduate studies in mathematics education, and in educational research. Her research interests are in the social, cultural and political dimensions of mathematics education theory and practice.

Chapter 2
In Conversation with Alan Bishop

Philip Clarkson

> Doing a graduate psychology course with Jerome Bruner switched me on. I thought to myself, we should be doing more of this stuff (research) in education, and in mathematics education. Gee! You know! Why are just psychologists doing this stuff? Soooo I took on various tutoring jobs just to check out some things. I tutored at a mental hospital. I taught and then tutored in schools in a black part of Boston in a program that Harvard ran with gifted black kids. I also taught in 'normal' classes in middle years. This really got me interested in research on teachers in the classroom.
>
> (Bishop reflecting on his time in Boston in the mid 1960s)

Alan was born in 1937, just before the Second World War commenced. His father was a mathematics teacher, who progressed to be a foundation principal of a new Grammar School in London. Alan's mother was a seamstress, who – not unusual for that time – concentrated on making a home for her husband and only child. One of the great joys of the family was music. His father played the violin for public performance in a trio, and his mother played the cello. Both gave Alan much active encouragement to develop his own musicality.

Alan sat for his 11 plus examination and scored enough to go to the University College School in London, a public school linked, originally, to London University. At school he chose to take a lot of mathematics and science, a lot of music and sport, all of which he has continued with throughout his life. Towards the end of secondary school, Alan successfully auditioned and subsequently played the bassoon for 2 years in the National Youth Orchestra. Clearly he had a wonderful, although for a young man, a difficult decision to make in those final years of schooling: would he concentrate on his music or mathematics? Taking the advice of a visiting musician from Holland, "Do you really want to enjoy your music? Then stay an amateur",

---

This chapter is mainly based on a number of conservations I had with Alan Bishop during April and May of 2008. But my conversation with Alan started with a brief question to him at a seminar he gave at Monash University in 1977. It continues through to today, in many and various locations including on golf courses, although those times should happen more regularly. Clearly the assertions and interpretations in this chapter are mine, although the dates and events have been checked with Alan.

P. Clarkson
Faculty of Education, Australian Catholic University, Fitzroy VIC 3065, Australia
e-mail: p.clarkson@patrick.acu.edu.au

P. Clarkson, N. Presmeg (eds.), *Critical Issues in Mathematics Education*,

Alan choose to continue his studies in mathematics, with music and of course sport as his second level studies.

At the conclusion of his secondary education in 1956, Alan chose to complete 2 years of national service. He entered the air force and spent most of that time as an air-radar fitter, which essentially meant trouble shooting the huge analogue computers then in use for navigation. This was Alan's first introduction to computers, and since this was 20 years or more before computer technology became widely available in society, he was considerably ahead of the game. On completing national service he presented himself for an interview at Southampton University, a normal part of the selection process. During the 30 minute interview, the Professor of Physics was far more interested in learning what Alan knew about computers, regarding his application for selection as a mere formality.

Alan had chosen to apply for Southampton since while concentrating on mathematics in his program, there would also be opportunity for music and sport as well. During his first year of study, he had the great fortune to meet up with Jenny, a talented linguist. They subsequently married, and still are supporting each other. His tutor turned out to be Bill Cockcroft, well known later for writing the Cockcroft Report in 1982, which advised the British government on strategies for revamping school mathematics. Interestingly it was just as much their common interest in jazz that sealed the beginning of a long friendship between Bill and Alan.

The notion of becoming a teacher had formed for Alan in his senior years in secondary school. He chose to pursue this interest by moving to Loughborough College on graduation from Southampton, since there he could undertake a 1 year Diploma in Education, not just for mathematics teaching but also in Physical Education. Alan was still in contact with Bill Cockcroft who suggested on the completion of his Diploma that he should apply for scholarships that would allow him to study in the United States, and incidentally get to know something of the interesting curriculum moves being made there with the so called "new math". Alan did win a scholarship through the Ford Foundation, so he and Jenny, now married, were off to Harvard University in the United States to complete an MA in Teaching. Although the scholarship was for 1 year, they stretched it out for 2 years, supplementing the scholarship monies with tutoring. They managed to stay for a third year by taking on full time school teaching in a local high school. Hence while taking classes with the likes of Jerome Bruner, Alan was teaching the new School Mathematics Study Group (SMSG) mathematics in high school, a wonderful preparation for his then glimmering idea of becoming a researcher in education. This glimmer of an idea is captured by the statement from Alan at the head of this chapter. It was at Harvard he started to see the possibility, and the excitement that can be generated, of doing good research.

Heading back to England after their stay in the United States, Alan rejected various school teaching jobs at top public schools, some of whom were teaching the new School Mathematics Project (SMP) mathematics curriculum, which would have ensured him a stable and well provided professional life. He was clearly well qualified for such jobs. But he rejected these lucrative offers, preferring instead to pursue this dream of researching in education. Hence he applied for and was appointed to a full time research fellow position at University of Hull working with Professor Frank Land. Unbeknown to Alan, Bill Cockcroft had moved to Hull, taking up the position of Dean of Science and Warden of one of the University Halls. Alan was delighted to take up the offer to be Deputy Warden to Cockcroft for his first 2 years at Hull. Apart from anything else, it provided him and Jenny with a free flat in which to live.

Land's 4 year project on which Alan was to work was centred on visualisation and the impact of this on mathematics learning. Although the project was very much in the psychology mould of doing research, nevertheless it was a project that was being carried out from within education. It was this subtle change that had excited Alan's interest at Harvard. Here at last he was starting to act out the idea. The project was basically assessing secondary school students on a range of visualisation and spatial ability measures, and on a number of attitude scales to do with mathematics. The students were also asked which primary schools they attended. At that time the primary education these secondary students had experienced in mathematics, formed a naturalistic but classic design for a research study. By ascertaining which primary school they had attended, the secondary students could be grouped into one of three groups: those who had completed their primary mathematics learning with the use of Cuisenaire materials; those who had used material devised by Dienes such as his MAB blocks and his logic blocks; and a third group who had experienced a traditional textbook resourced program. Interestingly those students who had used the various block materials in primary school, either Cuisenaire or Dienes materials, did much better on the spatial ability and visualisation tests, and had a much better attitude to geometry. The crucial aspect however of the study was later seen to be that the apparatus that the students had used in primary school was developed to help teach number concepts, not geometrical concepts, nor spatial abilities, nor visualisation. However it was in geometry that the real impact was made: this result seems obvious today, but in those days it was not so. These notions clearly linked with ideas that Alan had come across in the classes he had attended given by Bruner some years earlier. For Alan a real interest in visualisation and indeed spatial abilit*ies* of children grew, and this interest actively engaged him for the next 15 years or more. More comment is made on this focus in Section 3 of this volume.

At the conclusion of the project, and the completion of his doctoral studies in Hull, Alan moved to Cambridge University to take up a lectureship in the Faculty of Education that lasted for the next 23 years. He notes that he was regarded as an unusual appointment, because he did not come with the then normal 15 years or more of school teaching experience. Fortunately Richard Whitfield had gained an appointment in science education in this Faculty just before Alan's appointment. Whitfield also came with a research background rather than many years of school teaching experience. Interestingly Whitfield had been 1 year behind Alan at the same secondary school. Hence it is no surprise that once Alan had accepted the offer of an appointment, he and Whitfield joined forces to try to enliven the Faculty with a research program of their own.

The key to their project was to focus on the teacher in the classroom. Alan comments that then there were psychologists of various hues interested in studying the learner, often in "controlled" conditions out of the classroom, but gradually more and more working with the learners in the normal classroom situation. There were also curriculum colleagues more interested in *the* mathematics, thinking through what topics should be taught, in what order they should be taught, and since the break with the ossified traditional curriculum had been made, what resources could be brought in to help students learn. Many of the curriculum workers started to become aware of the psychologists and their findings on learning. But very few researchers were prepared to focus on the teacher in this mix.

The other critical ingredient that made this type of research possible at Cambridge was that they had access to video tape and video recorders. The video equipment was located in a suit of rooms in the Engineering Department. Hence bookings for it and relocation of students from their normal classrooms became a necessity. But nevertheless this apparatus gave the possibility of recording teachers teaching in situ, and then later replaying the recording and stopping the action at critical points to ask what became Alan's central question; "What might the teacher do next?" In listing possibilities of action before knowing what actually did occur, discussing them, and then evaluating these possible actions, Alan found a very powerful way to engage both practicing and beginning teachers in analysing their own and other's teaching. Hence this aspect of his research became known as the "teacher decision-making" phase. This became the enduring focus for Alan throughout his research career. In one way or another he has been asking, "And how will the good experienced (not the ideal) teacher teach the mathematics?"

As Borko, Roberts and Shavelson note in their chapter (this volume), the research on teacher decision-making did not take root in England to spawn an enduring research agenda. They go on to examine what then happened in the USA. However the echoing legacy of this research in England was not recorded in the research literature. In many tutorial rooms, both in England and parts of Australia used for pre service programs, video recordings of teachers are still being used in the way that was thought of in Cambridge in the early 1970s, the aim being to foster in inexperienced teachers, the ways of doing that experienced teachers just seem to know is correct for this moment and context. More comment is made on these research activities in Section 2 of this volume.

Clearly "doing research as educationists" was a novel idea at Cambridge at that time, as it was up to the early 1970s in Australia and elsewhere. Bishop and Whitfield were challenging a very fixed idea. It was all right for other disciplines to research learning, teaching and indeed all aspects of education. But those who practiced education as a craft really had no role as researchers. That notion seems quite quaint today.

During his time at Cambridge his engagement with a broad range of activities and people grew considerably, so by the time he moved to Monash University in 1992 he was a well known international academic with a rounded research pedigree. At Cambridge he was active in various ways within mathematics education in England, becoming a frequent speaker and convenor of workshops. He was active in various professional associations, including the Association of Mathematics Teachers (at one point Chair), the Mathematics Association (President for some years), and the

British Society for Research into Learning of Mathematics. One incident is instructive concerning his involvement with such associations. Alan tells of his attempts, alone and with others, to try to integrate the various professional mathematics associations during the 1980s, but to no avail. His concern was to have a strong united front, as mathematics education, as well as education generally, came under ever increasing pressure during the Thatcher years. To hear him speak of this time is to sense a deep regret that he and colleagues had not been able to make more headway on this political agenda.

However, working with individual teachers and small groups of teachers Alan always found profitable and exciting. He recounts a story of events that happened after he gave a talk for the Association of Mathematics Teachers on research in the early 1970s. Someone asked him at the conclusion whether ordinary teachers could engage in research themselves. Alan replied that essentially yes, although there were some protocols and procedures with which one should become familiar, and work within. He was then challenged directly after the talk by a small group of teachers who wanted to get going with some of their own research. From this interaction a small informal group of teachers grew, who did continue to engage in research in their own schools on their own teaching, with Alan as a mentor. The group included people like Geoff Giles, Kath Cross, and Bob Jeffreys. It began in 1972, developing a small but interesting series of studies using what would today be called action research.

His work gradually broadened on to the international scene during the 1970s. Part of this was through the people he had opportunity to meet. For example, the beginning of a long friendship, as well as opportunity for a rich academic partnership began on meeting Jeremy Kilpatrick for the first time at an invited working group in France in 1971 (see Kilpatrick's chapter 14, this volume).

These opportunities expanded when Alan, with others, developed and then began to teach an M Phil research degree program in mathematics education in the early 1980s at Cambridge, and also at about the same time began supervision of doctoral students. To comment on this today seems to be noting not much out of the ordinary, but it was then quite different. The earlier battles for engaging directly in research within education were starting to bear fruit, but even so there was still the lingering notion that practice was the normal and perhaps only aim of education, with research in education to be conducted by other more qualified social scientists rather than educationists. This meant another interesting difference, compared to the environment of today. Then there was much less pressure for tertiary education staff to have a coterie of research students. Alan notes that from time to time he would advise potential candidates to enroll elsewhere when he knew that they would be supervised by someone who had a deep interest in their particular set of research questions, rather than "grabbing" all candidates that came one's way, which is a tendency for some staff today. This mutual trust of colleagues across universities within Britain also helped meld the small but growing community of mathematics education tertiary staff into a very active supportive research group.

In taking these steps of engaging with teaching in research programs, Alan was brought into contact with colleagues from a number of countries. His first two doctoral students were Lloyd Dawe from Australia, and Norma Presmeg then from South Africa. The variety of students who enrolled in the 1 year M.Phil. program is also impressive: many have gone on to hold various positions in their own national professional education associations, as well as on the international stage. For example, Fou-Lai Lin, who was already a highly qualified mathematician and highly placed in the research administration in Taiwan, enrolled in the M.Phil. as his ideas turned to mathematics education. From the early days there was also Bill Higginson from Canada, and Renuka Vithal and Chris Breen from South Africa.

Alan also became active in international organisations. He attended the first International Congress on Mathematics Education (ICME) in 1969, and has since convened various groups for these conferences through the years. He was a founding member and co-director for 5 years of BACOMET (Basic Components of Mathematics Education for Teachers), an invitational international and hence multicultural research group that began in 1980 and continued to meet for more than 10 years. At times Alan held various positions in the International Group for the Psychology of Mathematics Education (PME) including being a member of the International Committee.

An important event that typified his work within these organisations concerned the year that PME was to meet in London during the mid 1980s. This was the time that world attention had finally turned to the apartheid question in South Africa. In line with a boycott of all things South African, there was a move to ban South African academics from attending the PME conference that year. After much arguing, the ban on the South African attendance was lifted, although the question was raised at the annual general meeting of the organisation. At Alan's suggestion, PME decided from then on not to ban attendance at the conference of any identifiable group of mathematics educators, even if such a ban could be seen as support of an acceptable political stance. Rather PME should find ways to support the attendance at its conferences of colleagues who are disadvantaged because of political situations, and such like. Putting this notion into action was another matter. An approach to UNESCO through Ed Jacobson by Alan to fund the publication of a book proved fruitful: the profit from the book was directed to PME. These monies became the founding amount for what has become the PME Skemp Fund, which continues to support the travel of colleagues who otherwise would not be able to attend PME conferences.

One of the mathematics educators that was influential in Alan's thinking was Hans Freudenthal. Freudenthal had founded what became one of the important international research journals in mathematics education, the *Educational Studies in Mathematics*. Alan was invited to become the second editor of this journal in the late 1970s (see Clements' chapter 7, this volume, for more discussion). He remains an advisory editor to this journal. This began for Alan a long association with the Kluwer Academic (now Springer) publishing house. In 1980 he founded and became the series editor for their Mathematics Education Library book series, a most highly regarded series that is still attracting authors. Within this series first in 1996, and again in 2003, two important two-volume handbooks were published that canvassed the state of mathematics education research worldwide.

However the most significant event that occurred during his time at Cambridge was in 1977. During the previous year Glen Lean from the Papua New Guinea University of Technology had visited Alan in Cambridge wishing to discuss with him the spatial abilities research that Alan had been involved with for 10 years or more. Glen's aim was to elicit support for the university students he was teaching who seemed to have great difficulty in mastering and understanding the geometry in the first year mathematics they had to study. Glen left with a parting invitation to Alan to visit sometime. Glen's visit certainly intrigued Alan. As it happened, Alan was planning to undertake a year of sabbatical through the 1977 academic year. An invitation had arrived from Professor Peter Fensham to spend some time at Monash University to work with Ken Clements. There was also an invitation to go to University of Georgia at Athens, USA, to link up with Jeremy Kilpatrick. Hence a year long round the world trip was planned for the family (by then Jenny and Alan had two sons) starting with 3 months in Papua New Guinea, then moving south to spend 5 months at Monash in Melbourne, Australia, and then finally travelling across the Pacific to spend time at the University of Georgia. It was the 3 months in Papua New Guinea that made the difference.

> "He changed" Ah ha! Yes he did.
>
> (Alan commenting on the first paragraph of Section 4 Introduction, this volume)

Ken Clements comments in his chapter (this volume) on the aftermath of Alan's Papua New Guinea visit in some length. This visit refocused Alan's interests in mathematics education away from his work with spatial abilities on to work with the impact of the social, cultural and political aspects on the teaching of mathematics. It seems, however, that this was not the first time that Alan had considered these other factors (in the traditional research way of thinking), or aspects of the educational environment, to say it a different way. A diagram first used by Bishop and Whitfield in the early 1970s, and reproduced in this volume by Borko et al. as Fig. 4.1 (see the introduction to Section 2, this volume), clearly has rectangles that suggest that during the 1970s Alan was well aware that the social, cultural and political aspects were important in understanding how teachers teach. His own experiences of school teaching in deliberately varied environments in Boston in the mid 1960s also alerted him to their individual and collective importance. A somewhat different experience in 1969 had also given Alan pause for thought. This concerned cross-cultural issues and forewarned him in part of the intricacies in trying to understand what was happening in such contexts. This experience was a keynote address at the first ICME conference given by Professor Hugh Philps from Australia, who reflected on his research conducted in Papua New Guinea. Philps' discussion of cross-cultural issues, which were mainly anchored in his Piagetian psychological studies with school students learning mathematics, fascinated Alan at the time. He spent some time talking with Philps at that conference. But even given these precursor experiences, it was his own experiences while living in Papua New Guinea that transformed Alan's thinking. No longer for him were the social, cultural and political issues of some importance; they became *the* important issues with which he needed to try and come to grips, as far as teaching mathematics was concerned.

Clearly Alan's concentration on these concerns can be seen in the headings used for the last four sections of this volume. The ways he chose to be involved with various professional groups noted above also indicates his new refocusing on these issues. His thinking was also stimulated by the small but engaged group of full time international students who came to Cambridge to enroll in the 1 year M.Phil. program that Alan started (see above), and the increasing numbers of doctoral students, again many from overseas. Within such a multicultural group, with most of the members already having substantial experience in education, Alan was able to test many of his own ideas as he sought to push himself into thinking through the implications of the political, cultural and social issues that impinged on mathematics teaching.

The key output from these years of reflection emerged as two books. The first is one of the most referenced volumes on mathematics education research, *Mathematical enculturation: A cultural perspective on mathematics education* (1988). Its sequel, which many do not realize is such, was the much later edited book by Abreu, Bishop, and Presmeg; *Transitions between contexts of mathematical practices* (2002). A plan that Alan had formed in the early 1980s, prompted by his Papua New Guinea experiences, was to develop two books, one on enculturation and another on acculturation. He was going to start with acculturation, but turned from that, being undecided on just how best to deal with the core notion, since he had never had to experience it directly. He then turned his whole attention to what enculturation means for mathematics teaching. Norma Presmeg, in her biographical notes in Chapter 1 (this volume), briefly comments on being a sounding board in the mid 1980s for Alan's ideas as the book came to fruition. One is not sure whether having lived in Australia for some 6–7 years, Alan finally felt he had some experience of acculturation, and hence was in a position himself to explore the long delayed second part of this writing program. Whether or not this is so, he interestingly had come to a way of breaking his blockage on this issue. Rather than deal with the idea front on by himself as he had with enculturation, he chose to think through the nuances of the idea, with a group of colleagues, using notions of transitions and indeed conflicts between cultures.

As noted above, Alan moved to Monash University, Australia, in 1992. This was not an easy move. It meant leaving their two grown boys back in England, and an aging parent. However the idea of growing old and crusty in an English establishment university was not the way Alan wished to finish his academic career. The move to Monash was attractive. It did mean promotion to a professorship, something that is not always a given in a place where one has been a long time. Both he and Jenny had enjoyed their extended stay back in 1977, and on subsequent visits had been made most welcome. He felt there were staff in the Faculty with whom he could easily form a working relationship. By this time Ken Clements had left Monash, and was on the staff at the Geelong Campus of Deakin University, a rural city about an hour's drive from Melbourne. Another interesting connection was that Glen Lean, who had inspired Alan's first visit to Papua New Guinea, was by then also on the staff at Deakin.

During his years as a paid staff member at Monash, from which he officially retired in 2002, Alan was heavily involved in the administration of the Faculty. He avoided the role of Dean with skilful footwork, but had different roles as Associate Dean, at various times, for Research, for International Affairs, and then as Deputy Dean, as well as being Head of the Mathematics, Science and Technology Group within the Faculty for some years. This of course meant membership and chairing of various Faculty and University committees. The time devoted to such increased through the 1990s as Monash, like universities elsewhere, moved totally into the age of performitivity and the attendant "need" for documenting everyone's activity to the nth degree, so that the organisation could work within a so-called "culture of evidence". Needless to say, much time was taken away from the core work of a high profile academic.

An early project that Alan worked on soon after arriving in Australia was to initiate the planning for an international regional conference through the agency of the International Commission on Mathematical Instruction (ICMI). This notion of ICMI supporting initiatives in particular regions of the world was not new, but certainly none had been contemplated for the South East Asian/Pacific region. However support was not always forthcoming from the Australians. In fact few in Australia had active involvement with the ICMI organisation, although they were regular attendees at the International Congress on Mathematical Education (ICME) four yearly conferences. Indeed when beginning the organisation for what eventuated as the 1995 ICME Regional Conference, no one in Melbourne was quite sure who was the Australian delegate to ICMI. Although such connections had been built for and during the 1984 ICME Conference held in Adelaide, Australia, 8 years later, interest in being actively engaged with this world wide organisation for many had waned. Hence those who had promoted the 1984 conference still held positions, even though lines of responsibility for action back to the mathematics education community were by then decidedly blurred. Alan's initiative inadvertently stirred up quite some angst. However the conference itself, although not as well attended as was hoped, still proved to be a success and cemented many connections between colleagues in Australia and overseas.

Although Alan continued writing on issues that he had started to think about before moving to Australia, he initiated two crucial research decisions. First he returned to the notions of what acculturation means in the mathematics education context. As noted above, this initiative finally produced an edited book, a conclusion to his original speculation some 20 years earlier on enculturation and acculturation. The other decision was to concentrate on values. This was not something new in Alan's thinking. The term appears in the diagram he and Whitfield used to conceptualize their ideas on teacher decision-making in the early 1970s. He had also begun to write quite explicitly on values by the early 1990s. By the mid 1990s Alan was ready to actively push the door of the classroom open again, and see what impact teachers' values had on their decision-making in the act of teaching. Thus began the Values and Mathematics Project (more comment on this is made in the Section 7, this volume). An Australian Research Council (ARC) Grant funded the original project. A subsequent project was also funded by a second ARC grant, but this time Alan joined with science education colleagues at Monash to broaden the scope of the investigation; an interesting turn of events which is reminiscent of his work with Whitfield, a science educator.

After a brief time in Melbourne, Alan linked with the local regional association of mathematics teachers, Mathematical Association of Victoria (MAV), for whom he had previously given seminars and a keynote presentation at their annual conference in 1977 (see Chapter 14, this volume). In this way he connected again with teachers who had been so much the centre of his research. He was a member of the MAV policy committee for some years. He also worked with the national professional group, the Australian Association of Mathematics Teachers (AAMT), to direct a project called Excellence in Mathematics Teaching. This was a joint project between Monash University and the teachers' association, and was funded by another ARC grant, with additional funding from various state government Ministries of Education. The main outcome was the development of a fully researched and trialled program that was aimed at senior mathematics teachers. The program led the teachers through some recent and relevant research, looked at some leadership issues pertinent for a mathematics coordinator, and also importantly included an emphasis on teaching skills. The teaching skills were not just discussed, but teachers were asked to view and analyse teaching episodes captured on video using the technique Alan had pioneered years before, as well as having some of their own teaching in their classroom observed and critiqued by others.

On coming to Monash, Alan had to take over the supervision of some research students who had been left without supervision with the retirement of other senior staff. However it was not long before additional local students and some from overseas were under Alan's supervision. His extensive travelling program helped this process. Mirroring his efforts at developing a group ethos among students at Cambridge, it was not long before monthly late afternoon seminars became the norm. The nucleus of these seminars were always Alan's research students, but also in attendance were often students being supervised by others in the Faculty, other interested staff colleagues both at Monash and from elsewhere, overseas or other visitors to the Faculty, and any research assistants employed to work on one or other projects then current. Clearly these categories were not always discrete, often including research students employed as research assistants, and staff members from elsewhere undertaking doctoral studies with Alan.

Among the overseas students that did come to work with Alan were some from Papua New Guinea: Wilfred Kaleva now Associate Dean of Education at University of Goroka, PNG, and Francis Kari. Such connections also enabled return visits to PNG from time to time, which both Alan and Jenny thoroughly enjoyed.

A bitter-sweet moment arrived mid way through the 1990s. Glen Lean, who had been the instigation of his first visit to Papua New Guinea, by then had developed into a lifelong friend. Glen became over the years another close and trusted critic (in the best sense of that word) for Alan's thinking about cultural impacts on mathematics. At the same time Alan became Glen's doctoral supervisor. Although Glen began his studies on the original issue that had led to his seeking Alan out (spatial abilities), after some years the study changed to a study of the mathematical systems embedded in the 800 plus languages spoken in Papua New Guinea. Glen was never good at consistently writing for his doctorate, and it must be one of the longest (timewise) doctoral studies ever completed. However, when it was finally finished, the four volume study, a cross between anthropology and mathematics education, preserves number systems and their analysis that are now dying out through lack of use, as the western system of education takes a real hold in that country. The thesis was finished after Glen had completed 21 years teaching in PNG, and joined the staff of Deakin University. However by the time of his graduation, specially arranged in Melbourne with the attendance of the Vice Chancellor of the PNG University of Technology, Glen had only months to live. Thus ended a lively, deep and thoughtful academic friendship.[1]

As Alan's time of retirement from the Monash academic staff approached at the end of 2002, his then current and past research students grouped together to nominate him for the University's Excellence in Research Supervision medal. He was subsequently awarded the medal at a graduation ceremony. On retirement, the University also granted him the accolade of Emeritus Professor, as recognition of his high quality contributions to the University across the areas of research, teaching, and in other ways.

In preparing to write this chapter I asked authors of the chapters contained in this volume what questions they would ask Alan if they were doing what I was about to do. Some of those questions and reflections have been embedded into the narrative above. However two remain with which it seems fitting to end. One was from Ken Ruthven who wanted to know "Looking back on your career, when and where were the occasions and situations that you felt that there was good (or better) alignment between the concerns and interests of mathematics education researchers on the one hand, and mathematics teaching practitioners and professional leaders on the other? What can we learn from these occasions and situations that might help develop and sustain such alignment?" In canvassing this question with Alan the conversation turned to those times when events from outside seemed to force themselves on to the concerns of mathematics education at large. There was the scare in the west of the Sputnik launch by the Soviet Union, and the question of whether the west was falling behind. The "something that had to be done" was, in part, the improvement of mathematics and science in schools. This took on different forms in England and the USA, but few of the proposals began within the mathematics education community. Teachers, professional leaders and those in universities had to respond and they in most part did so in concert with each other. The same happened with the first big influx of non-English speaking migrants into our schools in the 1960s and 1970s. The emphasis here was on the obvious language issues, and again there was some coming together to find solutions for praxis. Interestingly we seem to be revisiting this issue, but now in a broader way with the recognition of the multicultural mix in

---

[1] Glen's study can be found on the web site of the Glen Lean Mathematics Education Research Centre of the University of Goroka, Papua New Guinea: http://www.uog.ac.pg/glec/index.htm.

our classrooms, not just the embedded issue of language. Another issue that had all players in England asking "What do you do?" was the political decision to develop comprehensive schools, and hence mixed ability classes became the norm. Another was the rise of electronic calculators and computers, which came to schools via the business world. It seems that on each of these occasions when change was imposed from outside, at these times disparate sections of our community looked to each other for mutual support to find a way, first, of coping, and then to build again good praxis. These are the times we know we don't know, and hence we get together. More's the pity it takes such occasions for us to come together. Hopefully one day we will go beyond guarding our own small patches of turf, and realise that we are actually playing on the same sporting field.

Wee Tiong Seah's question picks up a slightly different but perhaps broader issue: "What do you identify as the main barriers to educational change today? How can our colleagues in research rise to this challenge?" Our discussion of this question seemed to dovetail with that driven by Ken's question. In seeking to become a profession, we in education seem to be very good at the moment in finding or creating barriers among ourselves, so that we have an identity that distinguishes us from the rest. And once there is a barrier, it has to be defended. But although at times it is good to have a robust identity, this should not prevent the crossing of the barrier to gain greater insight into problems that present themselves to all educators, no matter what type of hue we have (or think we have). Team research and easily accessible forums, which enable us to continue to speak and listen with each other, are always needed for the community to cope with the changes that often originate from elsewhere. It certainly seems we are being placed in a parlous state, at least in Australia, where good research teams are being asked to compete against each other for access to government monies to deal with issues that are of concern to all.

It has been an interesting experience to write this chapter. Not all of Alan's brilliant ideas, once put to the test of the classroom reality, have always come through with flying colours. He has not always won his political battles, although giving it his best efforts. You cannot mistake him for a god of mathematics education, or maybe even a guru, although his family name might lead some to think he is at least on the correct trajectory for one or other of those titles (well maybe in some after life). But what you can say about Alan is that he recognised early in his career that research needed to play a role in education, and in particular in the investigation of better ways to teach mathematics. Part way through his career he took notice of his experiences and strayed further from the orthodox research road that many of his colleagues were treading. Through all of this, his contributions have made a difference to many throughout our worldwide community in making others think more deeply about their untested assumptions, and indeed what they believe and why. And we acknowledge him for that.

I finish with a comment made to me by probably the youngest author contributing to this volume.

> His ability to not just think outside the box, but to do so in ways that are anchored to established knowledge and understandings separates Alan from others. This certainly has made it easier for real connections to be made in practice and research.
>
> (Wee Tiong Seah)

This chapter does not end here. As noted in Chapter 1, each of the following six sections begins with a brief introduction. These together should be seen as a continuation of this chapter.

References

Abreu, G. de, Bishop, A. J., & Presmeg, N. C. (Eds.). (2002). *Transitions between contexts of mathematical practices*. Dordrecht: Kluwer Academic Publishers.

Bishop, A. J. (1988). *Mathematical enculturation: A cultural perspective on mathematics education*. Dordrecht: Kluwer Academic Publishers.

Bishop, A.J., Clements, M., Keitel, C., Kilpatrick, J., & Leung, F. (Eds.). (2003). *Second international handbook of mathematics education*. Dordrecht: Kluwer Academic Publishers.

Bishop, A.J., Clements, M., Keitel, C., Kilpatrick, J., & Laborde, C. (Eds.). (1996). *International handbook of mathematics education*. Dordrecht: Kluwer Academic Publishers.

Section II
Teacher Decision Making

In the key article that begins this section, Bishop very clearly positions himself as a researcher who cannot be seen as purely an objective observer of the actions that take place in the classroom. This is an important notion since he never deviates from this line throughout his contributions to research. The researcher for Bishop, at least in the education sphere, must be seen as one who is influenced, who has influence, and is influential to varying degrees, within research projects.

This journal article by Bishop was written after 6–7 years of work focusing on what became a continuing and critical notion in his research: it focuses squarely on the teacher. More specifically, the article describes what for Bishop was perhaps the key to teaching: teachers making decisions in the act of teaching. Furthermore, Bishop explores what it is that can influence the teacher's potential decision outcome.

The article is also reactive. Bishop wants to see teaching through the teacher's eyes, and invites us to do so as well. He argues that experienced teachers know a lot about making good decisions, even if they have little time to consider their options in the flow of classroom activity. Researchers would do well to understand exactly how they do what they do so well. This position was in opposition to the university researchers' pre-service program term "teaching method", with its implication that we (the researchers) know what makes good teaching, if only teachers would follow our lead. Bishop argued then, and continues to do so, that we as researchers are better placed to observe closely and listen to what teachers have to say, and work with them.

As noted in the second chapter of this volume, the diagram that Bishop (1972) had devised in trying to conceptualise the parameters that impinge on teachers when making decisions in the classroom, can be used as a reference point for exploring Bishop's continuing research over the following decades. This figure (reproduced below, and found also as Fig. 4.1 in Chapter 4 by Borko, Roberts, and Shavelson) will be referred to a number of times in this volume. For example, in one of the rectangles of the diagram is to be found "values", an issue Bishop explicitly returned to in the mid 1990s, some 20 years later (see Section 7 of this volume).

The invited contribution to this section begins with a discussion of the key Bishop article and makes reference to the first book that Bishop published (with Whitfield). The discussion then moves to look at research that was being conducted in the USA

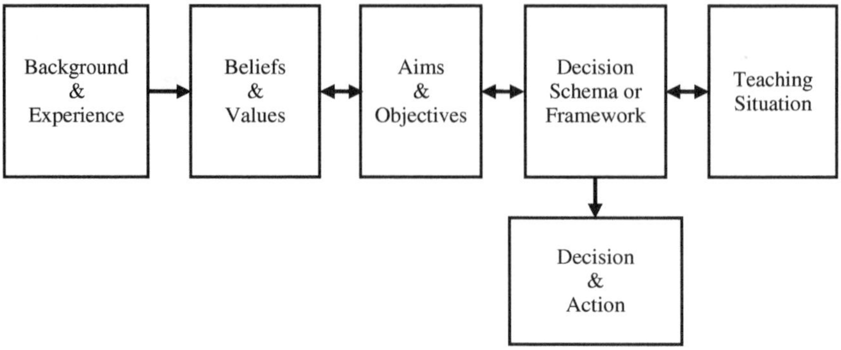

Fig. 1 Bishop and Whitfield's teacher decision-making framework
(adapted from Bishop & Whitfield, 1972, p. 6)

at the same time that Bishop was active on this issue. Noting that little more active research occurred in this area in England, the chapter traces the USA research that has focused on teachers making decisions. It notes that often as new researchers conceptualise what is happening in the classrooms of schools, there are frequently uncanny echoes of Bishop's often forgotten notions that he developed in the early 1970s.

Additional Bishop References Pertinent to This Issue

Bishop, A. J. (1972). Research and the teaching/theory interface. *Mathematics Teaching, 60*, September, 48–53.

Bishop, A. J. & Whitfield, R. (1972). *Situations in teaching.* London: McGraw Hill. [out of print]

Chapter 3
Decision-Making, the Intervening Variable

Alan J. Bishop

Teacher: Give me a fraction which lies between $\frac{1}{2}$ and $\frac{3}{4}$
Pupil: $\frac{2}{3}$
Teacher: How do you know that $\frac{2}{3}$ lies between $\frac{1}{2}$ and $\frac{3}{4}$?
Pupil: Because the 2 is between the 1 and the 3, and the 3 is between the 2 and the 4.

How would you deal with that response?

This example of momentary interchange in a classroom is presented to illustrate the heart of my research interest – immediate decision-making by teachers in the classroom. It is a subject which has concerned me, on and off, for the past seven years and having been invited to write an article for this journal I thought that it would be useful to describe some of the aspects of this subject which have been, and in some cases are still being explored.

It will help, I think, if I begin by outlining my personal research perspective, as no researcher can be objective and as there is no value to be gained from any pretended objectivity. I am motivated by my ignorance of how teachers are actually able to teach. Part of my ignorance is the shared ignorance which those of us who study teaching have and which is reflected in the relatively low-level descriptions and accounts of teaching which are found in books and in journals such as this. A second motivation which I have therefore is to encourage others in the pursuit of a deeper understanding of the teaching process. The task is far too great for any one individual alone. A third motivation is to do whatever I can to improve the quality of teaching. This means that I do not consider myself to be a 'neutral' researcher, and that, like anyone else, I have views about the criteria implied by any statements like 'improving the quality of teaching'.

I have described elsewhere (Bishop, 1972) my views about the links between research and the improvement of teaching, through the mediation of theory (in particular, the teacher's own theories) but I would like here to describe why the notion of decision-making is in my view such an important one, in this context.

Some years ago I was doing research on the effectiveness of different teaching methods in mathematics. For me it was interesting and challenging, but when discussing it with teachers and with student teachers I found the construct 'teaching

Educational Studies in Mathematics 7(1976) 41–47. *All Rights Reserved*

methods' not to be very understandable nor therefore particularly helpful. It was not a good vehicle for improving the quality of teaching! What was wrong was that 'teaching methods' is a researchers' construct – I can visit various classes and watch several teachers in action and attempt to describe the similarities and differences in their teaching methods. But a teacher who never sees other people teaching can only acquire a very limited idea of what 'teaching methods' means. In particular it is extremely difficult for him to separate out his methods from the rest of him – personality, style, mannerisms, etc.

Decision-making on the other hand is immediately understandable by teachers. There may be some discussion as to how conscious the making of choices is, or how important some decisions are when compared with others but anyone who has taught knows what it is like to be faced with the range of possible choices for dealing with, for example, the incident which started this paper. Or take another, possibly simpler, incident. You ask a child a question, she doesn't answer. Do you persist with her or do you ask someone else? If the latter, whom do you ask? Five children have their hands up, the rest have them down. Four boys at the back aren't even paying attention. Who do you ask? How do you ask? Perhaps it would be better (easier) to give the answer yourself. But how will you know if they understand? Perhaps that child does know the answer but she's just too frightened to answer publicly in case its wrong. Come on, you must *do* something.

And so it continues, incident after incident, with each choice being made under time pressure, under consistency-pressure (because you must be consistent and fair, mustn't you?), and under status-pressure (because after all, you are the one in authority?). Is it any wonder that teaching practice is such a traumatic time for student-teachers, that teachers are mentally exhausted after a day's teaching, that many teachers break down with severe nervous strain, or that teachers develop a powerful resistance to the 'ivory tower' ideas of those who aren't seen to share their pressurized existence.

Decision-making is therefore an activity which seems to me to be at the heart of the teaching process. If I can discover how teachers go about making their decisions then I shall understand better how teachers are able to teach. If we know more about teacher's decision-making then perhaps we can begin to relate theories about objectives, intentions, children's attitudes, children's mathematical development to the actual process of teaching. And I think that we shall therefore be in a better position to improve the quality of teaching.

Those are some of the reasons why I think this research area is so important. What then have I been doing? To talk briefly about the techniques I have been using – I have been tape-recording lessons of certain mathematics teachers, occasionally video recording them as well, and discussing incidents from those lessons with the teachers and with other teachers. In particular I have been looking at incidents where the pupil, or pupils, have indicated that they don't understand something, by making an error in their work or in their discussion with the teacher, or by not being able to answer a teacher's question, or by asking a question themselves. I have explored this in situations where the teacher has been teaching the class as a whole, where the teacher has been working individually, with a child either alone or in the whole class context. Some of the teachers have been experienced and some were inexperienced student teachers.

In some of the lessons I was sitting in and recording, whilst in others the teacher had the tape recorder there without me. All the teachers were known to me, were aware of what I was doing and aware of the general area of my research although they differed in the amount of knowledge they had about the specific types of incidents which interested me.

One particular technique I have used (and also incidentally, one which I use in my job of teacher training) is to 'stop-the-action' i.e., stop the tape when an incident occurs (such as the one at the start of the paper) before we see what the teacher does about it, and ask "What would you do now?" This naturally leads on to other questions such as "Why choose to do that?" "What other choices are open to you?" etc. I have also lifted out of the tapes some of these 'frozen' incidents and written them down in order to explore them away from the constraints of that particular lesson, that particular class of children and that particular teacher. (Some of these are published in a book. See Bishop, 1972.)

What then have I learnt from these various activities? One of the first ideas relates to the fact that when presented with an incident an experienced teacher usually smiles a smile of recognition, will often refer to a similar incident which happened to him recently and then say how he usually deals with such incidents. Clearly not every incident will provoke that response but enough do to suggest that experienced teachers have developed their own ways of classifying and categorising incidents into 'types' of incidents. It would be interesting to know what agreement exists between experienced teachers in terms of their dimensions of classification, and to speculate on what one might do with such 'agreement' if it exists.

Teachers appear to develop strategies for dealing with incidents 'which work for them'. They seem to assess the effectiveness of what they do at any incident and use that assessment to increase or decrease the use of their strategy. It is interesting to contrast several teachers responses to the same incident. Some agreement usually appears but the extent of the disagreement is striking. The teachers themselves comment on this, particularly those who put forward a strategy which no one else in the group would use. Occasionally there is a desire to pursue the notion of the 'best' strategy for any one incident, which can occasionally put the researcher in an awkward situation! The range of possible choices open at any incident is something which often surprises them as well. Take this incident as an example. A ten-year-old child has come to you with a subtraction problem and wants to know if her answer is right. Here are some of the possible choices open to you!

Ask yourself, what would you do and why? What type of option would you choose and why? What inferences, if any, would you be prepared to make about a teacher who chose one option rather than any others? What inferences would the child make about that teacher, about mathematics, about learning? What do any theories of learning offer in terms of judging the potential value of any particular options?

This particular incident can show another aspect of choice. It is set in a one-to-one teaching context, where the teacher has been going around the class and comes

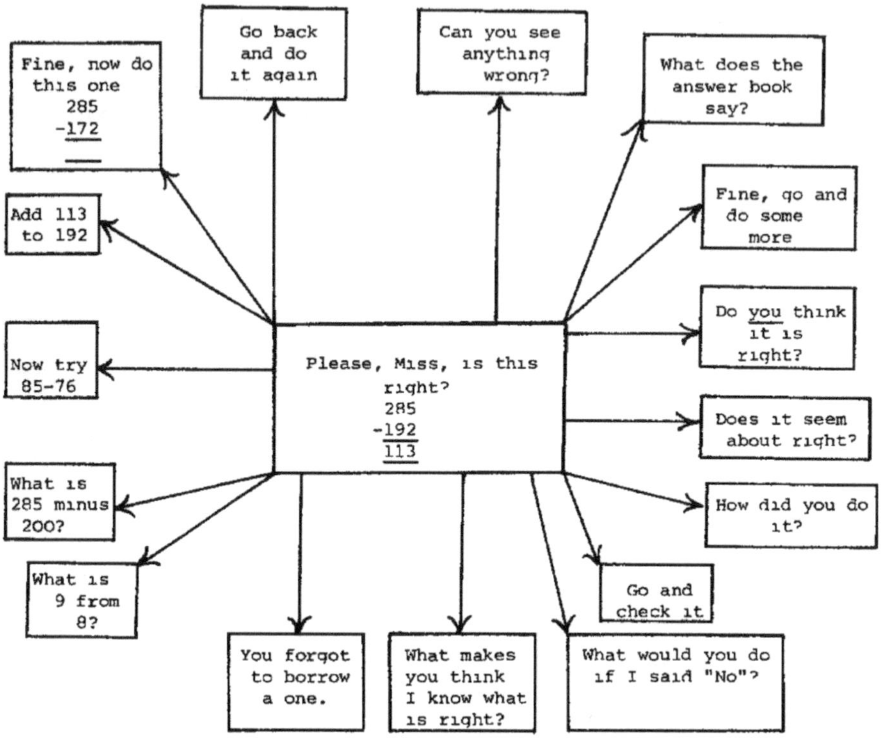

Fig. 3.1.

upon this child. What makes such an incident far more complex is if one meets it 'at the board' so to speak, with all the other children in the class watching and listening. It is to be discussed in the class? Should one seek answers from other children? Is it likely to be a common error? How can one best use that error for the good of the whole class? How will that child feel if her mistake is publicly exposed?

Teachers appear to be fairly consistent in their choices for dealing with such incidents. One teacher tends to point out that an error has been made, to help the child see the root of the error and then encourage the child to correct it. It has the flavour of a one-to-one strategy, with the other children merely observing. Another teacher usually repeats the child's erronious statement thereby opening the debate to the whole class. The fact that that is not usually done for a correct offering certainly cautions the children that when it is done the chances are that the statement is not correct. Another teacher enquires "Are you sure?" 'Is that right?" or repeats the statement questioningly (by the tone of voice). The fact that this teacher often does this with correct statements as well seems to encourage debate amongst the pupils.

The strategy of dealing with any contribution from the children without giving any hint of judgement seems to be a useful one in terms of the debate it encourages. It is not however, a strategy which is easy to employ successfully as the children seem to be fully aware of the implications of pauses which are longer than usual, of a facial reaction expressing surprise or annoyance, of inflections in the voice, and of the weight of emphasis on particular words. A strategy encouraging a variety of answers, from different children, each of which is recorded on the board seems easier to use and to be just as successful. The children are also aware of each other's competences and seem to develop a mental listing of how the teacher deals with contributions from particular children.

The teachers are of course also aware of different children's abilities and often use particular children as 'monitors' – if child A understands this point then the chances are that most of them will, if child B doesn't understand then probably most of them won't. Most teachers know which children they can call on to produce a right answer, which children would be embarrassed when asked an awkward question, which children will happily speculate publicly, and which children are likely to fall into a pre-arranged cognitive trap! Questions from the children are often classified in terms of the children – who genuinely is interested in the answer, who is trying to catch you out, who is simply wasting time, and who is trying to fool you into thinking that he's been listening. It must be clear by now, if it wasn't before, that in order to cope with a highly complex situation under quite severe pressure a teacher must develop consistent strategies which will allow him to survive. Learning to 'read' a classroom is clearly important, but one technique has emerged from this research which deserves wider publicity because of its implications. This is the technique which I call 'buying time'. It is a strategy which experienced teacher often use and which inexperienced students seem rarely to use (which has encouraged me to explore the technique with them). I will illustrate it with the incident at the start of the paper as that one happened to me, its on videotape and I can recall it vividly! What I did in dealing with the incident was: pause, smile, repeat the statement at length writing on the board, call it 'Jonathan's Law' in honour of the pupil who suggested it, pause, ask the other children "What does anyone else think about that?", pause and then write on the board a counter example. All-in-all I 'bought' for myself nearly 20 seconds of thinking time while I considered, is it true? How far does it go? Shall I open out the discussion? Shall we explore it together? Have I enough time? – No! How shall I 'close' it? Find a counter-example.

With the subtraction incident, also, some of the choices allow the teacher to buy time, more than others. This tactic, I suppose it too can be considered a choice within the decision-making process, is clearly used by teachers at certain points in their lessons. It is of course not always used to pre-empt a discussion, as I did above, but merely to provide a breathing space to allow the teacher to consider how he will deal with the incident. In some cases the teachers allowed the debate between pupils to go on unchecked for quite awhile. Meanwhile they were gaining all sorts of information while the interchanges were continuing independent of them – how many agree, disagree, what is the level of the discussion, what is the level of interest in the issue, what strategies of argument and persuasion are used by the children, etc. In fact bying-time often becomes 'acquiring more information' before deciding what to do – again as in the subtraction incident – does the teacher have enough information to know what to do, is it a simple accidental slip that the child has made? Is it a deeper misunderstanding? Of what – place value, subtraction, representation? Does she always make this error? Is it more helpful to this child to correct her mistake or to encourage her to develop mathematical independence from a teacher authority figure? If more information is required, what information should be sought, and how?

Disgressing slightly from decision-making for a moment, time-buying within lessons seems to be important for another reason. If the teacher is engaged in continual dialogue (or worse still monologue) throughout a lesson it is extremely difficult for him to 'stand back from the action'. Experienced teachers seem to recognize this need and create 'gaps' for themselves in the lesson.

In these gaps the teacher actively disengages himself from the learning process and occasionally seems to resent being re-engaged by, for example, a persistent child demanding attention. These disengagements seem to allow the teacher to relate what is currently happening to the longer-term picture, time spent on the topic, 'atmosphere' in the class, groupings within the class, work habits of particular children, etc.

Perhaps also they offer a period of mental respite from the demands of high-pressure interchange.

These then are some of the significant points which have so far emerged from the research. I am learning a great deal and now understand much more about (a) the complexity of teaching and (b) how teachers manage to cope with that complexity. I certainly believe that we can do a better job of initial training than we do at present, if only by making student teachers aware of the strategies for coping which experienced teachers use. I still feel ignorant, however, about the relationship between educational theory and the teaching process – how other people's ideas affect you when you are teaching. Do they suggest other choices which you hadn't previously considered? Do they offer criteria for judging the potential value of choices? The trouble is that these all sound too rational. Experienced teachers tend to opt for choices which work for *them*, in terms of their own personal criteria, indeed to rely on what appears to be a relatively limited routine response repertoire. It is significant that often they accuse me of glorifying their 'response system' by calling it decision-making. Only occasionally, at key points in the lesson, do they feel they are actively and consciously making decisions. But is this true of student teachers as well? It would seem that a beginner must be forced into making decisions more frequently simply because he has not yet had time for his routines to get established. But is this true?

And many more questions...

I hope then that this brief paper, written in a highly personal way, conveys the flavour, if not all the detail, of the research I am engaged in. I hope also that it may have persuaded a few more people that decision-making is a useful intervening variable, offering a window through which we may see more clearly the subtle interplay of abilities in that most complex of crafts, teaching. I hope finally that it has pointed out, if indeed it needed doing, that the development and improvement of mathematics teaching requires a great deal more awareness and understanding of the teaching process than is implied by the mere introduction of a new syllabus, new textbooks or a new examination.

Department of Education,
University of Cambridge

Bibliographical Notes

Bishop, A.J.: 'Research and the Teaching/Theory Interface' in *Mathematics Teaching*. No. 60, September, 1972.

Bishop, A.J. and Whitfield, R.C.: *Situations in Teaching,* McGraw-Hill, 1972.

For more ideas on the relationship between decisions and the development of 'open' attitudes and behaviours in children see:

Macdonald, J.P. and Zaret, E.: 'A Study of Openness in Classroom Interactions' in *Teaching – Vantage Points for Study* by Roland T. Hyman, published by J.B. Lippincott Company, 1968.

Also a paper by Bishop called 'Opportunities for Attitude Development Within Lessons' presented at the Nyiregyhaza Conference, Hungary in September 1975. Papers are to be published soon.

I would also like to record my thanks to my colleagues J. Sutcliffe and R.C. Whitfield for their continuing stimulus and encouragement.

Chapter 4
Teachers' Decision Making: from Alan J. Bishop to Today

Hilda Borko, Sarah A. Roberts and Richard Shavelson

Teacher: Give me a fraction which lies between 1/2 and 3/4
Pupil: 2/3
Teacher: How do you know that 2/3 lies between 1/2 and 3/4?
Pupil: Because the 2 is between the 1 and the 3, and the 3 is between the 2 and the 4

How would you deal with that response?

This example of momentary interchange in the classroom is presented to illustrate the heart of my research interest—immediate decision-making by teachers in the classroom.

(Bishop, 1976, p. 41, italics ours)

In the early 1970s, three scholars came to the realization, independently and almost simultaneously, that decision making was central to understanding and improving teaching. However, they took somewhat different routes in making this discovery and explored teachers' decision-making in somewhat different ways.

The first among them was Alan Bishop (along with colleague Richard Whitfield, 1972). They were grappling with two related issues. The first issue was that of the irrelevance of university teacher preparation: It was irrelevant to both the preparation of teachers and the practice of teaching. University teacher preparation placed too much emphasis on theory and too little on practical skills needed in the classroom. The second issue was the difficulty of providing teachers-in-training with experiences that would help them develop strategies for resolving problematic incidents as they occur in the classroom. For example, they needed safe, practical situations in which they could reflect upon different possible ways of handling problems before venturing into the classroom otherwise ill equipped for what they would encounter. Bishop and Whitfield said that placing teachers in classrooms without adequate

---

Authors listed in alphabetical order as each shared equally in writing this chapter.

H. Borko
School of Education, Stanford University, 485 Lasuen Mall, Stanford, CA 94305-3096, USA
e-mail: hildab@suse.stanford.edu

P. Clarkson, N. Presmeg (eds.), *Critical Issues in Mathematics Education*,

practical training was akin to putting pilots in the air before they had trained in a simulator, with the consequences being somewhat less dire.

They viewed teacher's decision making in a particular situation – the activity that "... seems to me to be at the heart of the teaching process" (Bishop, 1976, p. 42) – as a potential link between theory and practice. Bishop speculated, "If we know more about teachers' decision-making then perhaps we can begin to relate theories about objectives, intentions, children's attitudes, children's mathematical development to the actual process of teaching" (ibid.). To illustrate, he described being caught up by a pupil's justification for the answer that 2/3 fell in between 1/2 and 3/4. What should he do, in front of the class? The clock was ticking. As an experienced teacher, he employed a strategy that had worked for him and other teachers in the past – the strategy of "buying time." "What I did ... was: pause, smile, repeat the statement at length writing on the board, call it 'Jonathan's Law' in honour of the pupil who suggested it, pause, ask the other children 'What does anyone else think about that?', pause and then write on the board a counter example" (Bishop, 1976, p. 45).[1] He bought himself an estimated 20 seconds while he pondered whether Jonathan's Law was true, how general it was, what he should do to open discussion, whether there was enough time, and so on.

Experienced teachers like Bishop had come across a myriad of such situations and had classified them. Such classification meant that during the give-and-take of the classroom, a change in the teaching situation would most likely be recognized as a member of a class of like situations and an action would be readily at hand to respond – almost an automatic response based on prior decisions and their likely outcomes.

The other two who independently discovered decision making as central to teaching were Lee Shulman (with Arthur Elstein, 1975) and Richard Shavelson (1973). Shulman and Elstein, at Michigan State University, were in the process of completing a seminal study of physicians' clinical reasoning and decision making (Elstein, Shulman, & Sprafka, 1978). Intrigued by the potential relevance of some of their methods and findings to research on teaching (e.g., the dependency of problem-solving expertise on the clinician's mastery of a particular domain), Shulman brought these ideas to his program of research on teachers' problem-solving and decision-making.

Shavelson, working at Stanford University's Center for Research and Development in Teaching, had been inculcated into the teaching process-product paradigm where teachers' basic or technical skills of teaching were measured and correlated with student outcomes. Moreover, prospective teachers were taught isolated technical skills (about 20 at the time!) known to be correlated with student achievement, coached in deploying them ("micro-teaching"), and then observed and supervised in the classroom. He had been studying, at the time, the psychology and econometrics of human decision making and, like Shulman, viewed decision theory as a possible

---

[1] A counterexample would be 1/2, 2/3, and 5/8.

way of trying to understand and improve teaching. Shavelson (1973, pp. 392–393) reasoned that:

> [T]eachers are rational professionals who, like other professionals such as physicians, make judgments and carry out decisions in an uncertain, complex environment. . . . teachers behave rationally with respect to the simplified models of reality they construct. . . . teachers' behavior is guided by their thoughts, judgments and decisions.

While arriving at clinical reasoning and decision making along somewhat different paths, Shulman and Shavelson shared a common culture, that of emerging cognitive psychology, a culture that burst from the ashes of behaviorism almost phoenix like. Consequently they portrayed teachers' reasoning and decision-making as choices among alternative courses of action that were affected by their subjective estimates of the teaching situation, the actions available to them, and the likely outcomes all bounded by the limitations of human cognition in complex, ongoing teaching situations. To simplify the complexity, teachers developed "schemas" or frameworks for classifying and responding to the diverse situations they meet daily in classrooms. In doing so, however, they were open to predictable human errors of estimation and judgment and so the question arose as to how valid were the links in their schema.

So, three scholars – Bishop, Shulman and Shavelson – wandering along different paths discovered, independently, and at roughly the same time, decision making as central to teaching. They conceptualized teaching decisions as the fundamental link between complex, real-time teaching situations and practical actions in classrooms, postulating a cognitive framework or schema that underlies the link.

In this chapter, we expand upon Bishop's (and Whitfield's) ideas about teachers' decision making, the role of "situations" as simulations in enhancing the quality of teaching through reflective, deliberative debate and practice. We then trace teacher decision making research through the 1980s and 1990s concluding with the "revival" of the centrality of decision making in current research on teaching in the disciplines – a revival whose existence we recognize although this research does not label it as such and does not draw connections to Bishop's (or Shulman's or Shavelson's) work on clinical schema and decision making.

Bishop and Whitfield on Teacher Decision Making

Bishop and Whitfield (1972) believed that teacher decision making was ubiquitous and frequent. They distinguished pre- and within-lesson decisions as well as short- and long-term decisions (pp. 9–10). Pre-lesson decisions included objectives, content, method and materials. Within-lesson decisions included such things as implementation and/or modification of pre-lesson decisions, language level and logic, number and types of examples, error correction, motivating individual students, and discipline/social relations. That is, they typically dealt with learning (cognitive, attitudinal), relationships (with teacher, other pupils, other adults), and environments (materials, equipment, organization). Curiously, Bishop and Whitfield did not formalize post-teaching decisions in their framework, but such decisions were an integral part of the appraisal and revision of their postulated teaching appraisal system (see below).

Long term decisions primarily were those associated with pre-lesson decisions. These decisions could be made with deliberation and advice. Consequently, Bishop (1976, and Bishop & Whitfield, 1972) was most interested in those short-term, on-the-spot, within-lesson decisions that make teaching, in their words (1972, p. 2) so "harassing." And this is just where his work focused.

For Bishop and Whitfield, teachers based their decisions on prior experience, training, and practice. Each teacher developed individual decision-making frameworks or schema for making those decisions. In particular, Bishop and Whitfield believed that such frameworks could be developed for pre-service teachers first through *simulation* of teaching situations and then later through *practice in teaching situations*.

Teacher Decision-Making Framework

Teachers' mental frameworks or schema, according to Bishop and Whitfield, linked classroom situations to prior experience, values, and teaching goals in a classification/action table (matrix) which guided decisions and consequent action. More specifically, the framework enabled teachers to relate background information such as psychological theory as well as general life experiences, and especially educational experiences, to decisions about how to act in the everyday practical teaching situations facing them (see Fig. 4.1 adapted from Bishop & Whitfield, 1972, p. 6). This background information was filtered or interpreted through the individual's value system and aims/goals for the particular lesson. From talking extensively with teachers Bishop and Whitfield concluded that their "individual values do play a great part in their decision making, for example, their beliefs about the nature of persons and the nature of their subject material" (p. 6). In this way, Bishop and Whitfield viewed teachers' decisions as idiosyncratic.

To go back to the example at the beginning of the chapter, when Bishop was confronted with "Jonathan's Law," he had to decide, based on prior experience, his

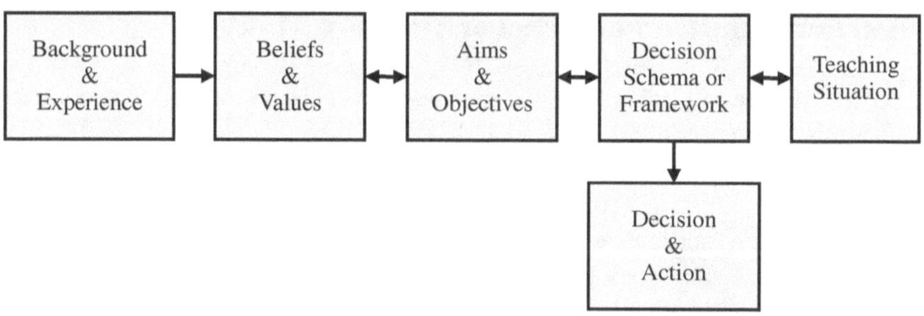

Fig. 4.1 Bishop and Whitfield's teacher decision-making framework (adapted from Bishop & Whitfield, 1972, p. 6)

beliefs and values, and the lesson goals, just how to handle the situation. Drawing on his decision schema, he decided to employ a well-know strategy – buying time – so he could think about the content ("Is it true? How far does it go?"), and the next pedagogical move ("Shall I open out the discussion? Shall we explore it together? Have I enough time?"). In the end he concluded that there was not sufficient time so he decided to "Find a counter example" (Bishop, 1976, p. 46).

Two systems underlie the development of these decision frameworks. One component is a classification system, which "... enables a teacher not only to recognize the similarities between situations, but also to determine the criteria which allow him to choose between the options open to him either consciously but often, with experience, tacitly or intuitively" (Bishop & Whitfield, 1972, p. 7). The second component, an appraisal system, assesses "... the success or otherwise of the particular action decided upon, and also ... [provides] information for selecting among choices ... in future situations" (Bishop & Whitfield, 1972, p. 7). Exactly how the appraisal takes place may be unknown to them, "... but able teachers are certainly aware of the quality of the decisions they have made" (p. 7).

Simulation of Situations in Teaching

Bishop and Whitfield reasoned that theory, including their theory of decision schema, plays itself out in practice when teachers act-in-context, appraise the results of those actions, and evolve a decision schema. With extensive experience, the decision table linking situations to courses of action becomes extensive and idiosyncratic. However, for the novice, situations in teaching present perplexing, demanding challenges. Novices must balance theory, their experience as a student, their beliefs about teaching and student learning, the aims of the lesson and possible courses of action for moving the lesson along. For novices to take off in their teaching air plane, so to speak, without crashing, thus requires considerable preparation and the construction, in a preliminary way, of a decision framework. Just as pilots are first trained in simulators, next fly under the guidance (dual control) of an instructor, and then fly solo, so should teachers learn to recognize paradigmatic teaching situations, identify possible courses of action, and make choices, by moving from simulation to guided practice to solo teaching. To be sure, this preparation with simulated teaching situations did not mean the novice could fly a classroom without careful attention, it just meant that the bumps encountered would be significantly lessened.

Bishop and Whitfield (1972) recognized that the use of simulation was new. Nevertheless, they believed that simulation was the "... area where the link between theory and practice can first be made, and where the decision framework to which we have referred can begin to be developed" (p. v). They proposed that short, taken-from-teaching teaching-situation scenarios should be "... all of a 'critical incident' nature ... [having] been compiled chiefly to give student teachers practice in thinking about appropriate on-the-spot decisions, under the guidance of their tutor" (p. v).

The teaching situations, then, serve as the basis for simulations. To this end, Bishop and Whitfield (1972) created a sort of taxonomy of teaching situations. One set of situations dealt with general types of teaching decisions such as (p. 26):

> It is half term and an unattractive 17 year old female pupil who has an unsettled home life has rung you up for the second evening in succession to say that she is depressed.
>
> *How would you deal with her problem?*

The other decision situations were drawn from teaching in two content areas: mathematics (Bishop) and science (Whitfield). The example at the outset of the chapter provides an illustration of a mathematics teaching situation.

Bishop and Whitfield envisioned teacher education students working either individually or in groups grappling with a particular teaching situation. They might, for example, encounter a situation where the teacher has introduced "matrices" in a mathematics class of 11 year olds and told them one reason they should be interested in the topic was because it had so many applications. "A rather spotty-faced boy, with glasses, sitting in the front row asks: 'Can you show us one, Sir?'" (Bishop & Whitfield, 1972, p. 19).

In analyzing the situation, Bishop and Whitfield suggest that perhaps the first thing the teacher-education student should do is decide whether the questioner was genuine. The answer to this question would influence subsequent decisions about what to say. "The teacher would first of all rely on his knowledge of this particular child and whether he was usually interested, bright, a show-off, or a teacher-baiter" (p. 19). They then go on to raise questions for teacher-education students to grapple with in the simulation (pp. 19–20):

1. List some of the alternatives open to you.
2. What are the criteria you must take into consideration here?
3. If you have no further information, what would be your best conservative strategy?
4. Suggest a good example.
5. Why might you choose to ignore the question?
6. If you decide to give him an example, what steps would you take to evaluate this decision?

Neither Bishop (1976) nor Bishop and Whitfield (1972) claimed to have "the answers" to the teaching situations. They noted the paucity of research on teaching situations. According to them, understanding how teachers make decisions and the impact of those decisions is ". . . unfortunately . . . something about which very little is known, and this is what research needs to focus on" (p. 4). Nevertheless, there are lessons to be learned from experienced practice, and Bishop (1976) went on to enumerate some of them.

> *What then have I learnt. . .?*
>
> (Bishop, 1976, p. 43)

Bishop asked this question rhetorically in his 1976 paper reflecting on his teacher decision making research. One of the first things he learned was that "... when presented with an incident an experienced teacher usually smiles a smile of recognition, will often refer to a similar incident which happened to him recently, and then say how he usually deals with such incidents" (p. 43). From this finding Bishop concluded, a propos the Teacher Decision Making Framework (Fig. 4.1), that experienced teachers had developed their own way of classifying and categorizing teaching situations into "'types' of incidents" (p. 43).

A second finding supported the conclusion that teachers' decision making schema were idiosyncratic in that teachers developed strategies that worked for them. "They seem to assess the effectiveness of what they do at any incident and use that assessment to increase or decrease the use of their strategy" (p. 43). Moreover, a third related finding was that "[e]xperienced teachers tend to opt for choices which work for them, in terms of their own personal criteria, indeed to rely on what appears to be a relatively limited routine response repertoire" (p. 47).

A fourth finding was that there are (a) multiple possible reasons for the errors students make and (b) multiple courses of action teachers can take in response to these errors. Teachers tend to be somewhat consistent internally in their choice for dealing with situations where students make errors. That is, one teacher will use the same or similar strategy over time. At the same time, however, teachers differ substantially from one another in the strategies they use for handling errors. "One teacher tends to point out that an error has been made, to help the child see the root of the error and then encourage the child to correct it" (p. 44). "Another teacher usually repeats the child's erroneous statement thereby opening the debate to the whole class" (p. 45). Yet "[a]nother teacher enquires 'Are you sure?' 'Is that right?' or repeats the statement questioningly" (p. 45). In all cases, these actions are taken without judging the child, although Bishop goes on to note the child does get the message, inevitably.

A fifth finding was that teachers are aware of different pupils' abilities and "... often use particular children as 'monitors' – if child A understands this point then the chances are that most of them will ..." (p. 45). Indeed, teachers have classified students just as they have classified teaching situations – students who are genuinely interested, a student "... who is trying to catch you out" (p. 45), a student representative of the lower portion of the class, and so on.

And a final finding was that only occasionally do experienced teachers feel they are consciously and actively making decisions, typically at key points in a lesson. In contrast, the "... beginner must be forced into making decisions more frequently simply because he has not yet had the time for his routines to get established" (p. 47).

Bishop's work, then, is firmly grounded in practice. From this grounding, he theorizes about the kinds of "cognitive structures" that teachers build up to tackle challenging classroom situations to make them manageable. One substantial contribution of this early work was its firm recognition of the importance of teaching situations or contexts and how they shape decisions and teaching actions. Bishop's approach to studying teachers' decision-making builds upon that recognition; as such, it stands in stark contrast to that taken by the cognitivists Shulman and

Shavelson. They started from theory and used that theory as a lens for understanding practice, as we shall see in the next section.

Shavelson on Teacher Decision Making

We now turn our attention to the cognitivists seeking to understand teachers' decision making. Because of the substantial overlap in Shulman's and Shavelson's ideas about teachers' decision making based on cognitive psychology, we focus on Shavelson's (1973, 1976; Borko, Cone, Russo, & Shavelson, 1979; Shavelson & Stern, 1981; for a general review, see Clark & Peterson, 1986). This said, there are also differences in their views; what follows reflects Shavelson's views and not necessarily Shulman's.

Teacher Decision Making Framework

Shavelson & Stern, (1981) conceived of teachers' decision making as encompassing their thoughts, judgments and decisions (Fig. 4.2). Within this conception, teachers are seen to integrate information about students, subject matter, and the school and classroom environment, filtering it through their beliefs and conceptions of the subject matter, so as to reach a judgment or decision on which their behavior is based (just as Bishop posited). Similarly to Bishop, Shavelson noted that experienced teachers had schema that linked teaching situations and teaching actions so that most "on-the-fly" decisions were made automatically; only when something went unexpectedly did these teachers become conscious of decision making. Finally, past and current judgments and decisions fed back their consequences to influence subsequent ones. This assessment and feedback process, then, is quite similar to that described by Bishop (1976).

Synopsis of Findings

To paraphrase Alan Bishop, "What have we learnt?" At least at a macro level, the findings from the cognitive approach to teachers' decision making (summarized in, for example, Shavelson & Stern, 1981; and Clark & Peterson, 1986) parallel Bishop's. The major difference lies in the cognitive theory that helps us organize and make sense of some of the findings.

For example, working in the field of cognitive psychology Shavelson found that in planning instruction, it turns out that teachers take into account student characteristics – ability, class participation, and problematic behavior – and that accounting depends on lesson goals and classroom, school and outside contexts. Teachers also take into account, to a lesser extent, students' sex, self-concept, and social competence, again depending upon the situation.

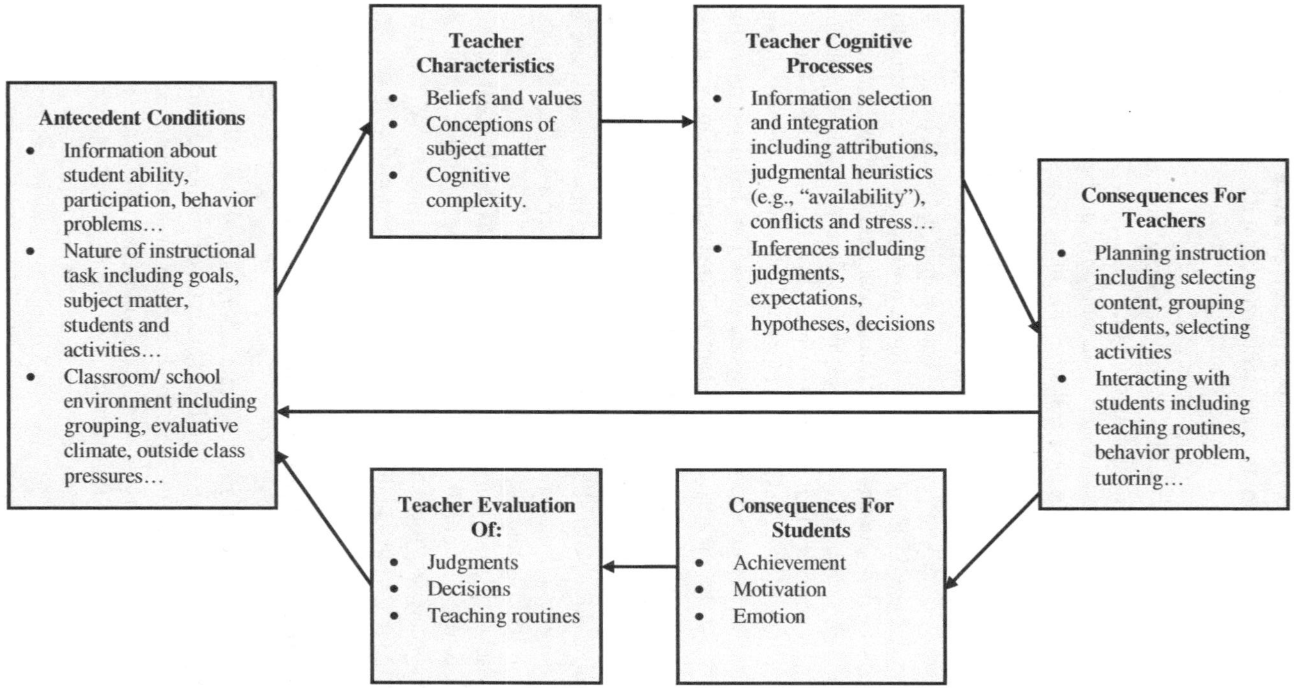

Fig. 4.2 Shavelson and Stern's teacher decision making framework (adapted from Shavelson & Stern, 1981)

With respect to instructional tasks, teachers focus on materials and activities; goals are less evident in their thoughts, most likely because they are embodied by the materials and activities. This said, there is some evidence that teachers' beliefs about the nature of learning (behavioral or constructive) influence their selection of tasks and materials, and the degree to which students are given responsibility in carrying out the tasks. Moreover, their conceptions of the subject matter (e.g., emphasis on comprehension or phonics in the teaching of reading) influence decisions such as whether or not to group students for instruction.

Shavelson and Shulman also found that, not only do beliefs about the nature of teaching and learning influence teachers' planning and subsequent interactive teaching, but teachers' judgments about students' knowledge, attitudes, and behavior and expectations of students' performance on class activities do as well. These judgments are typically based on short-cut strategies ("heuristics") that reduce information overload in a complex classroom. One such judgmental heuristic, the availability heuristic, refers to the ease with which instances can be brought to mind. When a description of a student matches the stereotype of a slow learner, for example, even if the description is unreliable, incomplete or outdated, teachers may predict with high certainty that the student is a slow learner and possess expectations about that student's performance in class. Another heuristic underlying teacher judgments is the anchoring and weighting heuristic. This heuristic states that initial judgments serve as an anchor to subsequent judgments based on observed performance. So, teachers who judge a student's performance to be low initially will tend to underestimate observed performance in their estimates of, say, student ability; and vice versa. It turns out that teachers, like all others, do use judgmental heuristics. Importantly, their estimates of student abilities in general are quite accurate, although this accuracy drops considerably when they are asked to estimate student performance on specific tasks.

Most of teacher planning focuses on creating tasks, and much of interactive teaching focuses on the smooth enactment of the task according to plan. Thus, once a task (materials and activities) has been formulated and sequenced, the formulation and sequencing operate as a plan, mental image, or script that the teacher carries out in the classroom. The task script guides the teacher's behavior during instruction – reducing information overload – until something goes unexpectedly. At this point, the teacher judges the criticalness of the situation and perhaps considers alternative courses of action.

Thus, like Bishop, Shavelson found that teachers build decision making schema that permit them to take into account their students' capacities, the nature of instructional tasks, and the teaching context. Such background information is filtered in a way consistent with their beliefs and values, to arrive at judgments about students and likely teaching situations that would move them toward instructional goals. In this way planning and interactive decision making are accomplished and enacted.

Here is where the work of Bishop and Whitfield came together with that of the cognitivists. We all view planning decisions as more deliberative than interactive decisions. And, we have come to understand that teachers have well-developed routines or scripts for teaching that link particular teaching situations created in planning to teaching actions taken during enactment. Such plans, scripts, and routines reduce the cognitive load that Bishop and Whitfield suspected overwhelmed novice teachers. Furthermore, we all believe that teachers idiosyncratically build up these schema over time and practice. The schema link background, values, goals and situations to teaching plans and actions.

Characterization of Teacher Cognition Underlying Decision Making

In his final writing on this topic, Shavelson (1986) cast teacher decision-making findings into a framework emerging from cognitive psychology. He noted that research on teacher thinking and decision-making had laid the foundation for a cognitive theory that might explain important aspects of teaching. He sketched one possible version of this theory of teacher cognition – schema theory – and showed how it might account for a variety of findings on teachers' planning and interactive decision making.

Drawing on work with Stern (Stern & Shavelson, 1983), Shavelson characterized teachers' cognition in the form of three schema – script schema, propositional schema, and scene schema. This characterization was similar in many ways to Bishop's, though set in the context of reading instruction (rather than mathematics or science): Teachers used common routines for teaching reading to low and high ability groups of students.

The script for the low group went something like this. The teacher typically: (1) provided a motivational exercise to start the lesson, (2) gave the format for the lesson, (3) began the lesson by reading aloud from the reader and then chose three of the students to read aloud in turn, (4) asked approximately ten questions about the factual content of the material, (5) reminded students of the format for their follow-up assignment, and (6) asked students to repeat the format correctly. In contrast, the teacher's script for the high reading group began by (1) discussing the story narrative, then (2) moved on to arguing about interpretation and causal structures, and (3) concluded by giving an overview of the next reading assignment with attention focused on story structure and causal arguments.

Shavelson formalized these low- and high-reading-group routines as a script schema – abstract information structures consisting of a set of expectations about the temporal order of events in routine procedures (for example, steps 1–6 in the low reading group; see Fig. 4.3). The script gets "instantiated" each day in the teacher's planning; while the categories in the script remain, their content changes. This formalization can predict what will happen in the low reading group from one day to the next, and from one teacher to the next. And when presented to the teacher as a "script," it may be used to change teacher behavior by changing the script.

Of course in the reading-group scenario the teacher brought to bear on the lesson her knowledge about her students, her goals for reading, the content (reading skills and meaning of the text read), possible instructional activities and materials, and her personal schema linking content and teaching strategies (Shulman's [1986b]

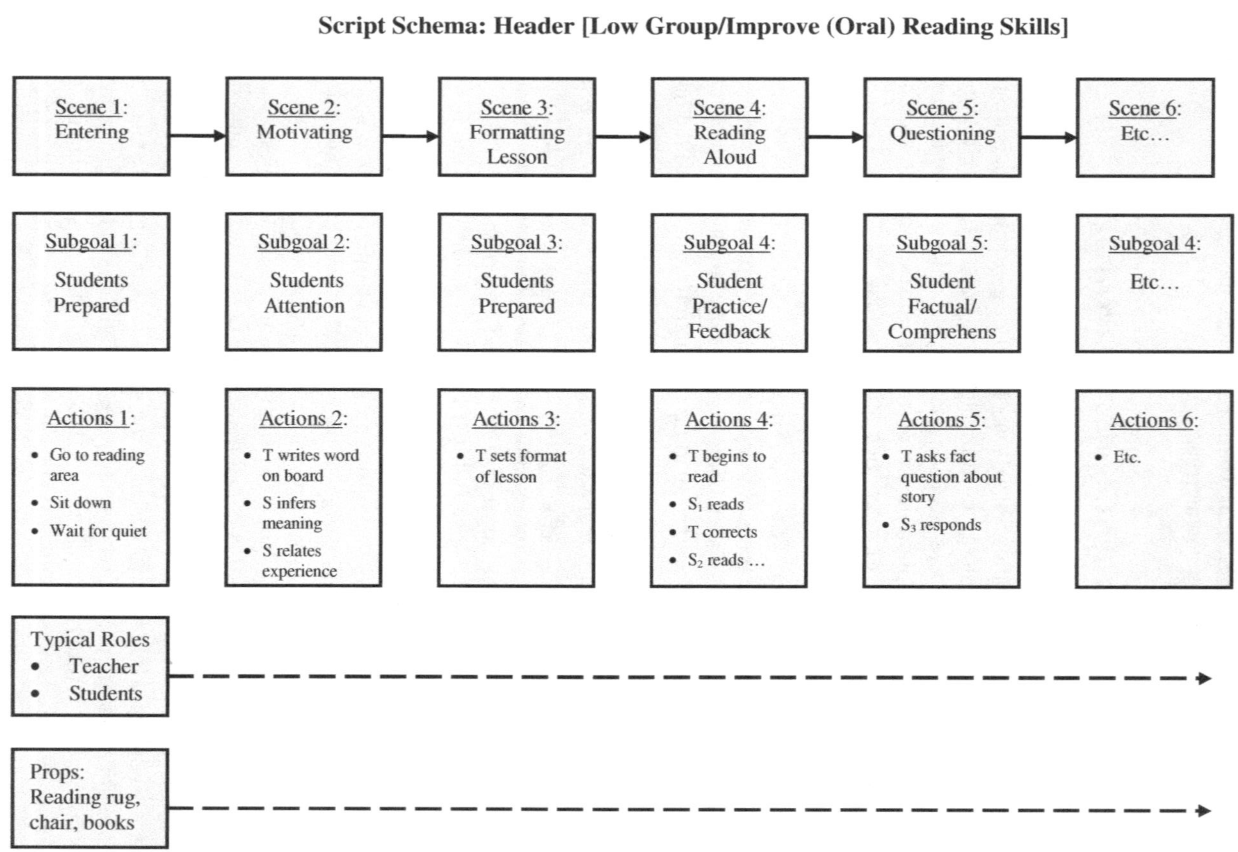

Fig. 4.3 Teacher script for teaching low reading group (adapted from Shavelson, 1986)

"pedagogical content knowledge"), in a manner consistent with class, school and outside contexts. Such knowledge and skills come in the form of propositional schema, abstract information structures that contain declarative (factual and conceptual) knowledge.

Finally, the classroom takes on a particular physical layout that affords and constrains activities. In the reading example, Shavelson and Stern focused on the "reading area" and its props (see Fig. 4.3). But it should be noted that, while 1/3 of the students were in the reading circle, the other 2/3 of students were working individually on a reading assignment (middle reading group) or project (high reading group). Part of a teacher's understanding of classrooms, then, is embodied in physical images. This knowledge is captured by scene schema. Such schema are spatial or topological in nature. They account for experienced teachers' recognition and flexible use of commonly occurring physical activity structures in classes.

Viewed from schema theory, the distinction between pre-active and interactive teaching begins to break down because planning involves "instantiating" a schema that gets enacted in the classroom. Unless the enactment does not go as planned, the teacher flows through the steps in her script. If the script is interrupted in a recognizable way, experienced teachers draw on routines they have developed to handle such "familiar" situations, as Bishop noted.

Indeed, Shavelson and Stern described the different scripts used with the low, middle, and high reading groups to their teacher. They then asked her to use the "high-group script" with the middle and low reading groups. They found that the teacher immediately understood what was being requested of her and changed teaching scripts accordingly. Students, in turn, changed their reading performance. Such scripts also offer promise for teacher education. Expert and novice teachers' schema can be distinguished and expert teachers' schema might be used as models in teacher preparation. This formalism seems to jibe closely with Bishop's notion of schema and how it guides classroom action. While each teacher's schema is organized, it is also idiosyncratic to the teacher as Bishop noted.

The potential for improving the quality of teaching seems evident from these early studies and theorizing about teacher decision making. Teaching situation simulations and teaching script and scene schema seem to provide valuable structure to teacher preparation. This said, however, there was little subsequent research on their potential. They seemed to fade away as other approaches gained in popularity. What might have been reasons for this turn of events?

Early Limitations of Teacher Decision Making Research

Unfortunately, little research at the time linked teachers' decisions to student learning or to what Bishop and Whitfield (1972) called teachers' "appraisal systems." This lack of focus on students – their class activities, cognitive processes and learning outcomes – led Shulman, (1986a, p. 24) to write, in his introductory chapter entitled "Paradigms and Research Programs in the Study of Teaching" in the third Handbook of Research on Teaching:

> Two serious problems beset the research program for the study of teacher cognitions. The first is the limited range of teaching activities about which teacher thoughts have been investigated. Other than the findings regarding teacher planning . . . little that is remarkable has emerged from the research studies. . .. The second . . . is the growing distance between the study of teacher cognition and those increasingly vigorous investigations of cognitive processes in pupils.

Returning to this point later in the chapter, Shulman continued, "Where the teacher cognition program has clearly fallen short is in the elucidation of teachers' cognitive understanding of subject matter content and the relationships between such understanding and the instruction teachers provide for students" (p. 25). With this statement he introduced "Knowledge Growth in Teaching," a major program of research on teachers' subject-matter knowledge upon which he and colleagues were embarking. In retrospect, we suspect that the statement also served as a premature obituary for the study of teacher clinical reasoning and decision making.

As we shall see in the next section, although some progress on teachers' decision making was made in the 1980s and early 1990s, programs of research on teachers' professional knowledge (including subject-matter knowledge) were becoming much more vibrant. Most recently, however, these two research areas are intersecting as a decision making lens, if not the terms "decision making" and "clinical reasoning," is being used to examine the professional knowledge teachers draw upon during teaching.

From Decision-Making to Knowledge: A Shift in Direction for Research on Teacher Cognition

In this section, we trace the shift in emphasis in research on teacher cognition from decision making in the early 1980s to professional knowledge in the late 1980s and 1990s.

Teachers' Planning and Interactive Decisions

Research on teachers' thinking in the 1980s and 1990s focused primarily on planning and interactive decision-making. Studies of planning – that component of teaching in which teachers formulate a course of action for carrying out instruction – examined topics such as reasons for planning, how teachers plan, and factors that affect planning. Interactive decision making refers to decisions teachers make while interacting with their students – "in-flight" or "real-time" decisions typically made without the luxury of time to reflect or to seek additional information. Researchers addressed a number of issues related to the nature, causes, and consequences of interactive decisions, and they developed several models of teachers' interactive decision-making. Much of this research used a cognitive psychological framework to analyze teachers' thinking. As such, it can be seen as more closely related to the work of Shavelson and Shulman, than to Bishop. Like Bishop, however, several of the researchers examined the role of teaching contexts in shaping teachers' decisions and actions.

Researchers during this time typically used methods such as policy-capturing and think-aloud to study teacher planning. In a typical policy-capturing study, a teacher is presented with a series of written descriptions of students, curriculum materials or teaching situations that vary on several dimensions, and is asked to make one or more judgments or decisions about each description. These judgments are used to generate mathematical models describing the relative weights the teacher attached to the features portrayed in the descriptions (i.e., the teacher's "decision policies"). Note that policy capturing is a sort of simulation, but not quite what Bishop had in mind as the simulations were built for research rather than educational purposes. Moreover, the situations were constructed to manipulate independent variables and not collected from real-world teaching situations as Bishop had done. With hindsight, some combination of approaches would have proven fruitful but this did not happen.

In a typical think-aloud study, the teacher is asked to verbalize all of his/her thoughts while making instructional decisions such as planning a lesson or judging curricular materials. These verbalizations are recorded, transcribed, and then analyzed to identify patterns in the content and processes of the teacher's thinking.

Most investigations of interactive decision-making used stimulated recall interviews, as did Bishop although he did not use the term, to elicit teachers' self-reports of their thoughts and decisions while working with students. The method of stimulated recall typically relies on audio- or video-recordings of teachers' lessons that are played back to "stimulate" their memories in interviews closely following the lessons. The cues or events that typically prompted teachers' interactive decisions were student cues such as disruptive behavior, unsatisfactory responses or work, and apparent lack of understanding. Teachers reported making real-time decisions about aspects of the instructional process such as questioning strategies, selection of student respondents, and selection of appropriate instructional representations and examples. (For a more extensive discussion of these methods see Clark & Peterson, 1986; Shavelson, Webb, & Burstein, 1986.)

Borko and colleagues, for example, investigated teachers' planning and interactive decisions about reading instruction. One series of studies used policy-capturing and think-aloud methods to examine teachers' strategies for grouping students for reading instruction. For both hypothetical students and children in their own classrooms, the teachers formed groups primarily on the basis of reading ability. Only when they could not easily make placement decisions for individual students solely on the basis of ability did they consider other factors such as class participation, motivation, work habits, and maturity (Borko & Niles, 1982, 1983, 1984).

Borko, Eisenhart and colleagues conducted a year-long ethnographic study of second-grade reading instruction in a rural Appalachian elementary school (Borko & Eisenhart, 1986, 1989; Borko, Eisenhart, Kello, & Vandett, 1984). They found that school-level and district-level policies constrained the teachers' decisions. Certain policies influenced teachers' interactive decisions, such as policies that required students to complete all textbooks, workbooks, and worksheets supplied by the required basal reading programs. Other policies influenced planning decisions such as administrative policies on class size, scheduling, grouping, and promotion and retention. For example, county and school guidelines specified when reading, language arts, and mathematics instruction would occur during the school day and how much time was allotted to each subject. Building administrators further influenced planning by assigning all students to reading groups within classrooms. Despite the clear influence of external forces, however, the teachers found opportunities for planning and interactive decision making within these constraints, which resulted in instructional programs that varied greatly across classrooms (Borko et al., 1984).

Planning and Interactive Thinking of Expert and Novice Teachers

Another line of research that flourished in the 1980s examined differences in the thinking and teaching of expert and novice teachers. David Berliner brought this research to the attention of the educational research community with his presidential address for the 1986 annual meeting of the American Educational Research Association, "In Pursuit of the Expert Pedagogue" (Berliner, 1986). Closely related to the work on teachers' interactive decision-making, research on pedagogical expertise reveals differences in the information expert and novice teachers attend to and the knowledge they draw upon while teaching.

The program of research conducted by Berliner and colleagues (Berliner, 1986, 1987) compared the ways of thinking of three groups of people: experienced/expert teachers (secondary mathematics and science teachers with at least 5 years of experience whose classroom teaching was judged to be excellent), novice teachers (highly rated student teachers and first-year mathematics and science teachers) and "postulants" (mathematicians and scientists who wanted to obtain certification for teaching but had no formal classroom teaching experience). In a series of studies examining participants' reactions to slides depicting classroom life (see Shavelson's scene schema), Berliner and colleagues found that experienced teachers differed from less experienced or inexperienced teachers in a number of ways. For example, they were more selective in the information they recalled. They also seemed to focus on information that had instructional significance, such as types of instructional strategies and classroom work arrangements. There were few clear patterns in the information that novices and postulants recalled, leading Berliner and colleagues to suggest that they did not seem to have a sense of what visual information was worth focusing on.

The experienced teachers in the research by Berliner and colleagues related the classroom scenes depicted in the slides to their own classrooms. Their responses were more interpretive than those of novices or postulants. Whereas the novices tended to describe surface characteristics of the classrooms and provide literal descriptions, the experts drew upon their personal knowledge to attribute meaning to the classroom scenes and explain the objects and events depicted. They also displayed more of a sense of typicality – spontaneously identifying slides that were out of order and attempting to make sense of the atypical. And, when provided with class records and information about students for a simulation activity in which they were to take over a class mid-year, the expert teachers spent very little time looking at these records. As they explained, the students were like others they knew, and they wanted to negotiate relationships with them without being influenced by information provided by others. Novices spent much more time on the task, apparently trying to make sense of all of the information provided. Postulants relied primarily on the text as a guide for where to begin instruction; for them, students were less important as sources of information than were class records indicating where the teacher had left off.

Considering their findings from a cognitive psychological perspective, Berliner (1987, pp. 75–76) concluded that the experienced teachers "have a fully developed student schemata, by means of which they operate." Furthermore, in comparison to the novices and postulants, "their perception is different, thus they remember different things. And their memory is organized differently. What they remember appears to be more functional." As Berliner noted, these findings and conclusions are compatible with Shavelson's work indicating "the enormously important role played by mental scripts ... in expert teachers' performance" (p. 72). Although he did not mention Bishop's research, they are also clearly related to his discussion of teachers' decision-making schema.

In Bishop's home country, England, James Calderhead (1981, 1983) conducted a series of studies comparing experienced teachers' and student teachers' general pedagogical knowledge. Participants in one study were asked to report on information they would need and strategies they would use when responding to common critical incidents in classrooms. The experienced teachers described situations similar to ones they considered typical of their own classrooms and commented on how they would normally deal with them. The student teachers distinguished fewer situations, focused on features such as time of day and importance of lessons, and responded more often with overall "blanket" reactions. A second study compared experienced teachers' and novice teachers' perceptions of students. In contrast to the novices, experienced teachers displayed a large quantity of knowledge about students and an awareness of the range of knowledge, skills, and problems to expect in the classroom. Similarly to Bishop as well as Berliner, Calderhead (1983) concluded from this program of research that, "Experienced teachers in a sense 'know' their new class even before they meet it" (p. 5).

Studies comparing the thinking and teaching of expert and novice mathematics teachers conducted by Leinhardt and colleagues (Leinhardt & Greeno, 1986; Leinhardt & Smith, 1985) and Borko and Livingston (1989; Livingston & Borko, 1990) reported similar findings and drew similar conclusions about the role of cognitive structures in pedagogical expertise. Patterns of differences in the thinking of expert and novice teachers identified in these research programs relate to the schema developed, as Bishop explains, through experience that teachers bring to bear in the analysis of classroom situations. When discussing their own or other teachers' classrooms, the experienced teachers were able to draw upon their personal knowledge of classrooms and students. They could therefore predict, identify, and deal with classroom problems more quickly and routinely than the novices. These patterns are clearly reminiscent of Bishop's finding that, "when presented with an incident an experienced teacher usually smiles a smile of recognition, will often refer to a similar incident which happened to him recently and then say how he usually deals with such incidents" (p. 43). The conclusions drawn by Berliner and Calderhead complement Bishop's suggestion that "experienced teachers have developed their own ways of classifying and categorizing incidents" (p. 43) and the schema described and studied by Shavelson. As Borko and Livingston (1989, p. 475) noted, "Many of the differences identified in the thinking of experts and novices acting in cognitively complex domains can be explained using the concepts of script, scene, and propositional structure." Like Bishop, Borko and Livingston considered implications for teacher education, suggesting that "developing these propositional structures and learning pedagogical reasoning skills are major components of learning to teach" (ibid.).

The Professional Knowledge Base of Teaching

One year before Berliner, Shulman also brought a developing line of research to the attention of the educational research community with his AERA Presidential Address, "Those Who Understand: Knowledge Growth in Teaching" (Shulman, 1986b). As in the Handbook chapter, Shulman introduced his "Knowledge Growth in Teaching" research program by connecting it to a limitation in existing research programs:

> My colleagues and I refer to the absence of focus on subject matter among the various research paradigms for the study of teaching as the "missing paradigm" problem. The consequences of this missing paradigm are serious, both for policy and for research.
>
> (p. 6)

Shulman's charge seemed to serve as a call to action. His research team, as well as several other educational scholars, turned their attention to teachers' professional knowledge and its relationship to classroom instruction. We address this body of work briefly. Although a key influence on the directions taken by research on teaching, it is less centrally related to Alan Bishop's legacy.

The team of researchers on the "Knowledge Growth in Teaching" project conducted a series of studies tracing the professional development of secondary teachers in their final year of teacher preparation and first year of full-time teaching. The project's theoretical framework included seven components of the professional knowledge base of teaching: knowledge of subject matter, pedagogical content knowledge, knowledge of other content, knowledge of curriculum, knowledge of learners, knowledge of educational aims, and general pedagogical knowledge (Wilson, Shulman, & Richert, 1987). Their own work focused primarily on two of these domains: knowledge of subject matter and pedagogical content knowledge. Without question, one of its key contributions is the construct, pedagogical content knowledge (PCK). As Shulman (1986b, p. 10) explained:

> Within the category of pedagogical content knowledge I include, for the most regularly taught topics in one's subject area, the most useful forms of representation of those ideas,

> the most powerful analogies, illustrations, examples, explanations, and demonstrations—in a word, the ways of representing and formulating the subject that makes it comprehensible to others.... Pedagogical content knowledge also includes an understanding of what makes the learning of specific topics easy or difficult: the conceptions and preconceptions that students of different ages and backgrounds bring with them to the learning of those most frequently taught topics and lessons....

Shulman and colleagues found that novice teachers begin to develop their PCK as a result of planning to teach and their experiences during instruction. For example, participants in their studies were able to generate more representations for each particular topic, and their representations reflected a deeper understanding of the topics. They also became better able to detect students' misconceptions and to correctly interpret their comments (Hashweh, 1985; Wilson et al., 1987)

Grossman (1990) expanded on Shulman's definition of pedagogical content knowledge, characterizing it as including four central components: (1) the teacher's overarching conception of the purposes for teaching a subject; (2) knowledge of students' understandings and potential misunderstandings of a subject area; (3) knowledge of curriculum and curricular materials; and (4) knowledge of strategies and representations for teaching particular topics. Several research programs conducted in the 1990s investigated the pedagogical content knowledge of novice and experienced teachers in each of these areas. Borko and Putnam (1996) reviewed much of this body of research, concluding that (p. 699):

> ...the pedagogical content knowledge of novice teachers is often insufficient for thoughtful and powerful teaching of subject matter content. And, although experienced teachers have generally acquired a good idea of pedagogical content knowledge, their knowledge often is not sufficient or appropriate for supporting teaching that emphasizes student understanding and flexible use of knowledge.. . .Novices have limited knowledge of subject-specific instructional strategies and representations, and of the understandings and thinking of their students about particular subject matter content. Experienced teachers typically have more knowledge of instructional strategies and of their students, but they often do not have appropriate knowledge and beliefs in the areas to support successful teaching for understanding.

"A Return to Bishop" Current and Future Research Connections

Many of the ideas that Alan Bishop put forth in his 1976 article are echoed in today's educational research. The terminology may not be the same, but the ideas are strikingly similar. In this section we explore those connections focusing, like Bishop, on mathematics education. While mathematics educators and researchers may not speak of "teacher decision making," they do talk about knowledge of mathematics for teaching, understanding and building upon student thinking, and using artifacts to inform and improve practice. All three of these ideas are highlighted in Bishop's work on teacher decision-making, and all look to be important topics for research and practice. While we discuss the three topics separately here, it is important to realize that in practice they often overlap one another and intersect in multiple points. For each, we discuss Bishop's thoughts on the topic, how current research ties into and extends his ideas, and what the anticipated and desirable future direction might be.

Mathematical Knowledge for Teaching

Mathematics Knowledge for Teaching (MKT) (Ball, Hill & Bass, 2005) has surfaced as an important topic in mathematics education within the past 10 years. Using Shulman's work as a starting point, Ball and colleagues have identified four types of knowledge needed for teaching mathematics:

1. *common content knowledge* – the type of mathematical knowledge and skill that is expected of any well-educated adult;
2. *specialized content knowledge* – knowledge and skills used to analyze and evaluate student errors, give mathematical explanations, and use mathematical representations;
3. *knowledge of mathematics and students* – an understanding of common student misconceptions, how to interpret incomplete student thoughts, how students are likely to respond to a mathematical task, and what students will find interesting and challenging; and
4. *knowledge of mathematics and teaching* – how to properly sequence content for instruction, recognize strengths and weaknesses of presenting material in certain ways, and how to respond to students' sometimes novel approaches.

Their research team is exploring a number of issues including the nature of mathematical knowledge for teaching, the role of MKT in classroom instruction, and its impact on student achievement (Ball & Bass, 2000; Ball, Thames, & Phelps, 2005; Hill, Rowan, & Ball, 2005).

Although typically not referenced in current work on knowledge of content for teaching, Alan Bishop suggested the importance of MKT in 1976. We discuss two examples within Bishop (1976) to illustrate. In both, Bishop suggests that there is more to teaching than just knowing a specific content area; teachers need to understand their content in ways that enable them to understand their students in order to help move student thinking forward.

In his own research, Bishop attempted to understand "how teachers go about making their decisions" (p. 42) by audio- and video-recording mathematics teachers' lessons and then discussing incidents from those lessons with the teachers and with other teachers. For example, he examined teachers' choices for dealing with incidents in which students revealed a lack of understanding through errors in their work, being unable to answer the teacher's question, or questions they asked of the teacher. In discussing this work Bishop (1976) focused on the consistency teachers demonstrated in their choices. Although he did not use the language of MKT, his discussion of patterns in teachers' choices clearly drew upon such knowledge: "One teacher tends to point out that an error has been made, to help the student see the root of that error and then encourage the child to correct it" (p. 44).

Bishop (1976) also used recordings of classroom incidents and techniques in teacher education. As he explained, he would stop the tape before viewers learned what decisions a teacher made and "ask 'What would you do now?' This naturally leads on to other questions such as 'Why choose to do that?' 'What other choices are open for you?' etc." (p. 43). These questions prompt the pre-service teacher to consider what the next steps might be and why a teacher would make that decision, while also considering additional options that might further student learning. In order to answer the questions, a teacher would need to draw upon her understanding of not only the mathematics she is teaching, but also her students and their mathematics growth trajectory.

A second type of incident Bishop studied that also illustrates the importance of MKT in teachers' decision-making, involves how a teacher responds to a student's query about whether her answer to a problem is correct. Using a child's question about a subtraction problem as an example, Bishop (1976) outlined the myriad choices available to a teacher in responding to that question. He then suggested a number of questions the teacher must consider in choosing a course of action:

> Ask yourself, what would you do and why? What type of option would you choose and why? What inferences, if any, would you be prepared to make about a teacher who chose one option rather than others? What inferences would the child make about the teacher, about mathematics, about learning? What do any theories of learning offer in terms of judging the potential value of any particular options?
>
> (p. 44)

Additionally, Bishop suggested that this incident would be much more complex if the teacher "meets it 'at the board'" (p. 44) rather than talking one-on-one with a student. During a whole-class activity, a teacher is on a larger stage for his or her students. The teacher is forced to make fairly instantaneous decisions about where to move a discussion and why. Bishop's questions delve into the many types of knowledge that a teacher must draw upon in order to be able to address her students' learning needs. Among other things, the teacher must be prepared with an understanding of the strengths and weaknesses of solution strategies, as well as common student errors.

Turning again to the 21st century, we find that several researchers are exploring similar types of incidents (student errors or lack of understanding, their queries about answers to problems) and raising similar questions about these incidents. Yet, for the most part, their focus (and language) has been on teachers' professional knowledge rather than their decision-making. Prominent among these scholars, Ball and Bass (2000) note, "managing these real situations demands a kind of deeply detailed knowledge of mathematics and the ability to use it in these very real contexts of practice" (p. 84). Answering the questions that Bishop (1976) posed in the example above would require a teacher to understand the mathematics of the subtraction problem, beyond the general understanding of the subtraction algorithm as Ball, Hill and Bass (2005) suggested.

In our own work, two of the authors (Borko and Roberts) have drawn upon this conceptualization of MKT to develop and study the Problem-Solving Cycle (PSC) – a professional development model for middle school mathematics teachers that provides them with opportunities to increase their professional knowledge and improve their instructional practices. As part of the Supporting the Transition from Arithmetic to Algebraic Reasoning (STAAR) project middle school mathematics teachers participated in a series of workshops that focused on both learning content knowledge and learning how to share that content with their students. In the first workshop, teachers completed rich mathematics problems that they would later be using with their own students. The two subsequent workshops in each series used videoclips from the teachers' lessons as a springboard for discussions of the students' reasoning and teachers' instructional practices. Although we did not draw upon Alan Bishop's work (at least, not explicitly or consciously) many of the incidents depicted in the videoclips were similar to those that he studied. In working through the problems, teachers were able to identify possible areas in which students might have difficulty. They also developed a better understanding of different solution strategies students might use and the strengths and weaknesses of those strategies. When teaching the problems, they started asking questions like the ones that Bishop (1976) proposed, and they demonstrated an understanding of mathematics beyond knowing the algorithms. Our analyses suggest that by the end of the project, participants had increased their professional knowledge in the domains described by Ball and colleagues (Ball, Thames, & Phelps, 2005).

There are many questions still to pursue when we consider mathematics knowledge for teaching. Much research will likely describe and measure teachers' knowledge of mathematics for teaching. Two key measurement tools recently developed by the Learning Mathematics for Teaching (LMT) Project at the University of Michigan are a paper and pencil test and a video coding instrument (Hill, Ball, & Schilling, 2004; LMT, 2006). Additional tools, such as microanalysis of video (Erickson, 1992), might provide a more in-depth look at teachers' MKT. Microanalysis could allow researchers to study the moment-to-moment work of a teacher as she works through the content and attempts to manage her classroom. For instance, we can imagine zooming in on a teacher's questioning strategies and her responses to students' questions. There are a number of nuances, such as the tone of a teacher's or student's voice and the looks on students faces, that help illustrate what is happening with the mathematics in a classroom and how a teacher is responding to it. This level of detail is not readily available through the other two measures. Using a tool like the LMT video coding scheme to frame what researchers should look for in microanalysis of classroom video, it might be possible to create a continuum of MKT that ranges from novice to expert for skills such as probing student thinking or explaining mathematical concepts. This continuum would help researchers to gauge areas where teachers have difficulty and possibly to create learning opportunities for teacher professional development and teacher education. For instance, schools of education and university mathematics departments could provide content courses for pre-service and in-service teachers that help them develop rich content knowledge as well as an understanding of how to use that content to respond to students' questions, to organize and create curriculum, and to manage classroom discussions. Clearly, moving in these directions would benefit greatly from explicit attention to Bishop's theoretical and empirical work on teachers' decision-making.

Understanding Student Thinking

Another burgeoning line of inquiry showing great promise for teacher education is the exploration of teachers' understanding of student thinking. Since the National Council of Teachers of Mathematics wrote the first and subsequent standards documents (NCTM, 1989, 1991, 1995, 2000), mathematics classrooms in the US have begun to transform considerably. Teachers are attempting to teach mathematics in ways that differ substantially from the mathematics instruction that they received as students. To teach in these ways, they must be able to facilitate discussions, delve into open-ended problems, and manage a classroom equitably (NCTM, 2000). It is no longer enough to grade students on whether they get a problem right or wrong. Instead teachers must be able to determine how a student arrived at a particular solution, to ask questions that help develop the students' mathematical thinking, and to engage students in understanding their peers' mathematical work. Key to accomplishing these tasks, they must have an understanding of their students' thinking in order to move the mathematics work along for the class and individual students. Bishop (1976) made a very similar point when he suggested that a teacher had to understand whether a student had made a simple accidental slip or had a deeper misunderstanding, the nature of the misunderstanding, and whether that student always made this error, before deciding whether to correct the error or to encourage the student to delve more deeply into the mathematics. A teacher's rich knowledge of mathematics for teaching – particularly knowledge of students and mathematics – is essential to this work.

Understanding student thinking requires teachers to dissect student comments during discussions and to analyze their written work. They must be able to follow the reasoning paths their students took, analyze errors in their written work, and ask the kinds of questions that enable them to gather more information about the work that both individual students and groups are doing. Once again, we see a foreshadowing of contemporary explorations of student thinking in Bishop's (1976) work on teacher decision-making. Bishop explained, "If I can discover how teachers go about making their decisions then I shall understand better how teachers are able to teach" (p. 42). As we noted earlier, much of this work focused on classroom incidents characterized by student errors or misunderstandings: "I have been looking at incidents where the pupil, or pupils, have indicated that they don't understand something, by making an error in their work or in their discussion with the teacher, or by not being able to answer a teacher's question, or by asking a question themselves" (p. 43). In such situations a teacher has a responsibility to evaluate why a student is making a particular error, why a student cannot answer a question, and to *decide* how to adjust the mathematics classroom work in light of those errors and questions. Although Bishop probed for teachers' decisions, implicit in his lines of questioning was attention to how teachers understood and classified these situations, and how they reacted as a result of understanding their pupils' paths.

Similarly, several contemporary programs of research are investigating teachers' understanding of student thinking and how their instruction builds on that understanding. Perhaps not surprisingly their explorations are framed by assumptions and questions about teachers' professional knowledge rather than decision making. Teachers need a rich and complex understanding of content and students to be able to understand students' thinking and appraise their work (Ball, 1997). To understand students' mathematical thinking and work, they draw from their specialized mathematics knowledge for teaching and knowledge of mathematics and students, as well as their more general knowledge of students. In addition to knowing how eighth grade students might approach a problem and misconceptions they might hold, the teacher must be aware of the more general factors that affect eighth graders. For instance, how can a teacher convince a student to explain her mathematical processes when the student is mostly thinking of the fight she had with her mother last night.

One of the most challenging tasks in understanding student thinking is that most often teachers must do this work in their moment-to-moment instruction. Often there is little time to analyze their students' work. Ball (1997) wrote:

> The challenges to figuring out what children know are great. Facing two and three dozen students at a time, teachers must make ongoing, usable, estimates about what individual students know. They must observe and listen, and they must interpret their observations, reach conclusions, and act.
>
> (pp. 806–807)

Again, we are reminded of Bishop's work.

Bishop (1976) suggested that it is possible to gain insight into the mathematics that students are doing (and what they understand about that mathematics) by allowing students to work while the teacher takes a seat in the audience. Focusing again on the incident in which he was attempting to understand how a student had solved a subtraction problem, one technique Bishop used was to ask other students to share what they thought of that student's solution strategy. Bishop explains this instructional move to his readers: during times when students' thinking is unclear a teacher might have to engage in "buying time" (Bishop, 1976, p. 45). For instance, she may do what he did: "pause, smile, repeat the statement [a student made]. . .[and] ask the children 'What does anyone else think about that?'" (p. 45). This approach allows the teacher to acquire more information about the student's mathematical thinking, how his or her thinking fits in with the work of the class, and how to move the class forward. Bishop later had other teachers try this technique. He noted that as the students were talking through their understanding of the problem, the teachers "were gaining all sorts of information while the interchanges were continuing independent of them" (p. 46). These strategies enable a teacher to step back from the work that students are doing, generally by way of asking students to explain their own thinking or that of a fellow student. They "allow the teacher to relate what is currently happening to the longer-term picture" (p. 46), so that they can see how the work they are doing fits in with the larger learning trajectories that students are working within.

Teachers can use the knowledge they gain from understanding student thinking to guide decision-making about the next steps of instruction. A teacher's understanding of her students' thinking and reasoning will help her determine which activity she should plan next, which piece of content would best support the development of students' foundational understanding, and which questions she should ask to help students develop an even richer understanding.

While many of the decisions that a teacher makes about students' thinking must occur in the moment in a classroom, they can develop their ability to understand student thinking and their decision-making skills in contexts outside the classroom. A professional community of mathematics teachers can provide the structure for teachers to examine student thinking. And videoclips of classroom incidents, used in ways similar to those Bishop described, can provide the context and substance for their collective inquiry. Two such examples of collaborative communities are described by Kazemi and Franke (2004) – who worked with teacher workgroups in an elementary school that were examining student thinking and classroom practice, and the STAAR Project (Jacobs et al., 2007) – where teachers focused explicitly on student thinking in the third and final workshop of the Problem Solving Cycle.

Building on early Cognitively-Guided Instruction work that laid out an "organized set of frameworks that delineated the key problems in the domain of mathematics and the strategies children would use to solve them" (Franke & Kazemi, 2001, p. 43), Kazemi and Franke (2004) organized and facilitated workgroups as places for teachers to share the mathematical work that was occurring in their classrooms. During some workgroup sessions teachers shared their students' methods for solving problems, and they often found that these methods were distinct from the ones they might have used themselves. Upon realizing this, the teachers' involvement in class and workgroups seemed to shift, to focus on making "...sense of and to detail their students' thinking" (p. 229) through the use of classroom artifacts and rich workgroup discussion. This exploration helped the teachers to develop a better understanding of their students' thinking, benchmarks in students' learning trajectories, and instructional trajectories to support students (Kazemi & Franke, 2004).

Similarly, the third workshop in the Problem-Solving Cycle focused primarily on student thinking, addressing topics such as how students explained their solution strategies, and their misconceptions or naïve conceptions (Jacobs et al., 2007). To foster these discussions, the facilitator selected videoclips and student work, often centering on a student's or students' novel approach to solving a problem. The teachers developed a better understanding of how and why students approached a problem in a particular way. This provided them possible questions to ask students next time this problem was used and the ability to possibly transfer their understanding of this context to others. Jacobs et al. note, "As they reflect, individually (in writing) and collaboratively (in small or whole group discussions), the teachers...consider how they might improve their instructional practices based on knowledge gained thus far."

Franke, Kazemi, Shih, Biagetti, and Battey (2005) noted that the importance of generative learning within professional development. Teachers must continue to learn about their students and use this to inform and transform their practice, independently of university research projects. One way to do this is by making the practice of examining student thinking a regular occurrence, during in-class interactions as well as within communities of professionals. However, it is not enough for teachers to simply examine student thinking, they also must use that information to inform decisions in their classrooms. For example, in the workgroups facilitated by Kazemi and Franke, teachers examined student thinking through investigation of student work and interviews with students, located students within a learning trajectory, and then used this information to connect students' understandings to their other knowledge structures (Franke & Kazemi, 2001). As Bishop suggested, we can also do a better job at initial teacher preparation by helping student teachers to become aware of these strategies that experienced teachers use.

Use of Artifacts for Teacher Education and Research

Artifacts of classroom practice, such as videotapes or audiotapes of classrooms, student work, teacher journals, lesson plans, and curricular materials are all tools for inquiry. Artifacts may come from a teacher's own classroom or from the classrooms of others. In either case, they allow teachers to situate the work of exploring and improving their teaching in practice, for example, by enabling them to "investigate what students are doing and thinking, and how instruction has been understood as classes unfold" (Ball & Cohen, 1999, p. 11). Video, initially used for microteaching, has recently become paramount in providing teachers with the opportunity to examine their own and others' practice in naturally-occurring classroom situations.

Bishop's use of video for teacher education in the early 1970s foreshadowed this more recent development. Bishop (1976) described how difficult it can be for teachers who do not have the opportunity to see other people teach to really know what teaching can entail: "a teacher who never sees other people teaching can only acquire a very limited idea of what 'teaching methods' means. In particular it is extremely difficult for him to separate out his methods from the rest of him" (p. 42). He used audio- and video-recordings in teacher education to discuss classroom incidents with student teachers. Sometimes he would "stop the action," pausing a tape and asking teachers such questions as, "What would you do now?" and "Why choose to do that?" to allow them to explore possible routes a teacher might take. Without such artifacts as video and audiotapes, these conversations would likely have been quite limited, as the artifacts grounded the conversations and allowed critical incidents to be seen and heard exactly as they occurred in the classroom.

There are a number of characteristics that make video especially useful as a tool for teacher education and professional development. It is cost effective, easy to edit and reassemble (Brophy, 2004; Seago, 2004), and provides a lasting record (Sherin, 2004). Video allows teachers to pause, replay, analyze, and reanalyze classroom situations, providing access to others' practices as well as their own (Seago, 2004). While it is powerful in and of itself, it is also easily used in conjunction with other artifacts, such as lesson plans and student work.

It is now easier than ever to record, edit, and use video, especially with the advent of digital technology. Recently, a number of professional development/research projects have taken advantage of these advances, examining the use of video with teachers in ways that not only "stop the action" as Bishop did, but go beyond.

For example, the STAAR project and Miriam Sherin's work with video clubs both use videos to facilitate teacher learning about content, student thinking, and lesson development within mathematics classrooms. Teachers in Sherin's video clubs met to discuss excerpts of videos from their own classrooms (Sherin & Han, 2004), while STAAR video-recorded members of the professional development group teaching the same lesson and then used selected video clips in subsequent workshops as springboards for exploring student thinking and the role of the teacher. These projects found that over the course of 2 years, the teachers began to focus more on issues related to teaching and learning mathematics and on student thinking, and less on describing their own teaching or the teacher in the video clip (Borko, Jacobs, Eiteljorg, & Pittman, 2008; Sherin & Han, 2004). The projects demonstrate the power of video in helping teachers to develop mathematics knowledge for teaching.

Use of video as a tool in research on teaching also was still in its infancy when Bishop (1976) wrote about his experiences studying teacher decision making. As the discussion of Sherin's video clubs and the STAAR Problem-Solving Cycle workshops indicates, several research teams are currently using video as a research tool to describe and analyze teacher learning, as well as a pedagogical tool to support that learning. Video can be particularly powerful in this regard when analyzed using instruments and methods grounded in theoretical perspectives on teacher learning. When Bishop wrote about teacher decision making in 1976, he expressed dissatisfaction with his (and the field's) understanding of the relationship between theory and practice. Commenting that, "I still feel ignorant, however, about the relationship between educational theory and the teaching process," he suggested that explanations of teachers' actions derived from theory "all sound too rational" (p. 47). In the final section of this chapter, we consider the divergent paths taken by programs of research on teacher decision making that were grounded in practice and those that were grounded in theory, and we suggest a possible convergence that builds on recent efforts to elaborate a practice-based theory of teacher learning (Ball & Cohen, 1999) and to design and study professional development that uses artifacts such as video to situate teacher learning in the practice of teaching (Putnam & Borko, 2000; Borko, 2004).

Concluding Comments

Thirty-five years ago Alan Bishop, with his colleague Richard Whitfield, conceived of teacher decision making as the nexus between *thought* and *action*. The relationship was reciprocal. Thought involved teaching schema – a sort of practical argument linking situation to action – that evolve idiosyncratically for each teacher out of practical experience with consequences. Action was the observed teaching practice. Decision making linked *thought* and *action*. Through simulation and then immersion in guided practice, pre-service teachers would build increasingly complex teaching schema thereby bringing relevance back to university-based teacher education.

Bishop's ideas were grounded firmly in practice. Teaching situations were developed from teachers' descriptions of incidents and their practices in those incidents. Situations provided the basis for simulations. And pre-service teachers would evolve their schema through dialogue with one another and under guidance of mentors about what alternative courses of action might be taken and what the likely consequences would be. Remarkably, out of this research strategy emerged a framework for teacher decision making that presaged advances in cognitive psychology and in conceptions of professional knowledge for teaching.

The work of Shulman and Shavelson, based on cognitive psychology, provided theoretical underpinnings for what Bishop had envisioned. Such theorizing moved the field ahead, stimulating substantial research on teacher cognition. However, to some extent this psychological theorizing and research moved away from the complexities of practice. Simulations were built for research, not teacher preparation, purposes. In contrast to Bishop, cognitive psychologists built simulations to test hypotheses that emerged out of theory. The good news is that substantial understanding was developed regarding the nature of teaching schema. The bad news is that, as Shulman recognized, the theoretical cognitive focus proved limiting, and student cognition and performance were left out of the mix. Here lies the origin of a premature obituary for research on teacher decision making.

The "new new thing" that replaced decision making in research on teaching is teacher learning. Teacher education (pre-service and in-service) and research have focused on the professional knowledge needed for teaching, and on ways to help teachers develop that knowledge. Moreover, teachers' understanding of student thinking has played a prominent role. This research is guided by a conception of the professional knowledge needed for teaching. It is also steeped in practice: understanding classroom situations and transactions between teacher and student that link content and pedagogical knowledge in the development of a pedagogical schema. But what seems to be absent is what Bishop was concerned about. The link between knowledge and action is unclear. Also lacking, once again (or still) is theory that connects knowledge to action *in practice*. Knowing does not mean effective doing.

Educational researchers have made substantial progress toward developing a theory of professional knowledge for teaching – a theory that describes what teachers need to know in order to teach effectively, as well as how that knowledge is manifested in classroom practice (Ball & Bass, 2000). And, importantly, initial research evidence indicates a positive link between professional knowledge for teaching and student outcomes (Hill et al., 2005). We have also begun to develop a practice-based theory of teacher learning – one that draws upon a theory of situated cognition and proposes using artifacts of practice (e.g., video) to situate the development of teachers' professional knowledge in classroom practice (Ball & Cohen, 1999; Putnam & Borko, 2000). Since Bishop's seminal piece, however, there has been almost no theoretical work linking teacher knowledge to classroom action. One notable exception is Alan Schoenfeld's (1998) research program to develop a theory of teaching-in-context. Like Bishop's decision-making framework, Schoenfeld's model attempts to link teachers' knowledge, beliefs, and goals to their decisions and actions in the classroom. It seems that a return to these concerns, with the advances in theory and research findings, may be appropriate. Teacher decision making may, once again, be a useful construct for the study of teaching linked to student outcomes.

References

Ball, D. L. (1997). What do students know? Facing challenges of distance, context, and desire to hear children. In B. J. Biddle, T. L. Good, & I. F. Goodson (Eds.), *International handbook of teachers and teaching* (Vol. 2, pp. 769–818). Dordrecht, The Netherlands: Kluwer.

Ball, D. L. & Bass, H. (2000). Interweaving content and pedagogy in teaching and learning to teach: Knowing and using mathematics. In J. Boaler (Ed.), *Multiple perspectives on the teaching and learning of mathematics* (pp. 83–104). Westport, CT: Ablex.

Ball, D. L. & Cohen, D. K. (1999). Developing practice, developing practitioners: Toward a practice-based theory of professional practice. In L. Darling-Hammond & G. Sykes (Eds.), *Teaching as the learning profession: Handbook of policy and practice* (pp. 3–32). San Francisco: Jossey-Bass.

Ball, D. L., Hill, H. C., & Bass, H. (2005). Knowing mathematics for teaching: Who knows mathematics well enough to teach third grade, and how can we decide? *American Educator, Fall*, 14–22, 43–46.

Ball, D. L., Thames, M. H., & Phelps, G. (2005, April). *Articulating domains of mathematical knowledge for teaching*. Paper presented at the annual meeting of the American Educational Research Association, Montreal, QC, Canada.

Berliner, D. C. (1986). In pursuit of the expert pedagogue. *Educational Researcher, 15*(7), 5–13.

Berliner, D. C. (1987). Ways of thinking about students and classrooms by more and less experienced teachers. In J. Calderhead (Ed.), *Exploring teachers' thinking* (pp. 60–83). London: Holt, Rinehart & Winston.

Bishop, A. J. (1976 July). Decision-making, the intervening variable. *Educational Studies in Mathematics, 7*(1/2), 41–47.

Bishop, A. J., & Whitfield, R. C. (1972). *Situations in teaching*. Maidenhead Berkshire, England: McGraw-Hill.

Borko, H. (2004). Professional development and teacher learning: Mapping the terrain. *Educational Researcher, 33*, 3–15.

Borko, H., Cone, R., Russo, N. A., & Shavelson, R. J. (1979). Teachers' decision making. In P. L. Peterson & H. Walberg (Eds.), *Research on teaching: Concepts, findings, and implications* (pp. 136–160). Berkeley: McCutchan Publishing Corp.

Borko, H. & Eisenhart, M. (1986). Students' conceptions of reading and their reading experiences in school. *The Elementary School Journal, 86*, 589–611.

Borko, H. & Eisenhart, M. (1989). Reading groups as literacy communities. In D. Bloome (Ed.), *Classrooms and Literacy* (pp. 107–132). New Jersey: Ablex.

Borko, H., Eisenhart, M., Kello, M., & Vandett, N. (1984). Teachers as decision makers versus technicians. In J. A. Niles & L. A. Harris (Eds.), *Thirty-third yearbook of the National Reading Conference*. New York: The National Reading Conference, Inc.

Borko, H., Jacobs, J., Eiteljorg, E., & Pittman, M. E. (2008). Video as a tool for fostering productive discussions in mathematics professional development. *Teaching and Teacher Education, 24*, 417–436.

Borko, H., & Livingston, C. (1989). Cognition and improvisation: Differences in mathematics instruction by expert and novice teachers. *American Educational Research Journal, 26*, 473–498.

Borko, H. & Niles, J. A. (1982). Factors contributing to teachers' judgments about students and decisions about grouping students for reading instruction. *Journal of Reading Behavior, 14*, 127–140.

Borko, H. & Niles, J. A. (1983). Teachers' cognitive processes in the formation of reading groups. In J. A. Niles & L. A. Harris (Eds.), *Thirty-second yearbook of the National Reading Conference*. New York: The National Reading Conference, Inc.

Borko, H. & Niles, J. A. (1984). Instructional planning. In J. Keefe & J. Jenkins (Eds.), *Instructional leadership handbook*. Virginia: National Association of Secondary School Principals.

Borko, H. & Putnam, R. (1996). Learning to teach. In D. Berliner & R. Calfee (Eds.), *Handbook of educational psychology* (pp. 673–708). New York: MacMillan.

Brophy, J. (Ed.). (2004). *Advances in research on teaching: Vol. 10. Using video in teacher education*. Oxford, UK: Elsevier.

Calderhead, J. (1981). A psychological approach to research on teachers' decision making. *British Educational Research Journal, 7*, 51–57.

Calderhead, J. (1983, April). *Research into teachers' and student teachers' cognitions: exploring the nature of classroom practice*. Paper presented at the annual meeting of the American Educational Research Association, Montreal, Canada.

Clark, C. M., & Peterson, P. L. (1986). Teachers' thought processes. In M. C. Wittrock (Ed.), *Handbook of research on teaching* (3rd ed.). Washington, DC: American Educational Research Association.

Elstein, A. S., Shulman, L. S., & Sprafka, S. A. (1978). *Medical problem solving: An analysis of clinical reasoning*. Cambridge, MA: Harvard University Press.

Erickson, F. (1992). Ethnographic microanalysis of interaction. In M. LeCompte, W. Millroy, & J. Preissle (Eds.), *The handbook of qualitative research in education* (pp. 201–225). New York: Academic Press.

Franke, M. L. & Kazemi, E. (2001). Teaching as learning within a community of practice: Characterizing generative growth. In T. Wood, B. Nelson, & J. Warfield (Eds.), *Beyond classical pedagogy in elementary mathematics: The nature of facilitative change* (pp. 47–74). Mahwah, NJ: Erlbaum.

Franke, M. L., Kazemi, E., Shih, J., Biagetti, S., & Battey, D. (2005). Changing teachers' professional work in mathematics: One school's journey. In T. A. Romberg, T. P. Carpenter, & F. Dremock (Eds.), *Understanding mathematics and science matters* (pp. 209–230). Mahwah, NJ: Erlbaum

Grossman, P. (1990). *The making of a teacher: Teacher knowledge and teacher education*. New York: Teachers College Press.

Hashweh, M. Z. (1985). *An exploratory study of teacher knowledge and teaching: The effects of science teachers' knowledge of subject-matter and their conceptions of learning on their teaching*. Unpublished doctoral dissertation, Stanford University, Stanford, CA.

Hill, H., Ball, D., & Schilling, S. (2004). Developing measures of teachers' mathematics knowledge for teaching. *Elementary School Journal, 105*, 11–30.

Hill, H. C., Rowan, B., & Ball, D. L. (2005). Effects of teachers' mathematical knowledge for teaching on student achievement. *American Educational Research Journal, 42*(2), 371–406.

Jacobs, J. K., Borko, H., Koellner, K., Schneider, C., Eiteljorg, E., Roberts, S. A. (2007). The Problem-Solving Cycle: A model of mathematics professional development. *Journal of Mathematics Education Leadership, 10*(1). 42–57.

Kazemi, E. & Franke, M. L. (2004). Teacher learning in mathematics: Using student work to promote collective inquiry. *Journal of Mathematics Teacher Education, 18*, 917–946.

Koellner, K., Schneider, C., Roberts, S. A., Jacobs, J., & Borko, H. (in press). Using the problem-solving cycle model of professional development to support novice mathematics instructional leaders. *AMTE Monograph V: Inquiry into Mathematics Teacher Education*.

Learning Mathematics for Teaching (2006). *A Coding rubric for Measuring the Quality of Mathematics in Instruction* (Technical Report LMT1.06). Ann Arbor, MI: University of Michigan, School of Education.

Leinhardt, G. & Greeno, J.G. (1986). The cognitive skill of teaching. *Journal of Educational Psychology, 78*, 75–95.

Leinhardt, G. & Smith, D. (1985). Expertise in mathematics instruction: subject matter knowledge. *Journal of Educational Psychology, 77*, 241–247.

Livingston, C. & Borko, H. (1989). Expert/novice differences in teaching: A cognitive analysis and implications for teacher education. *Journal of Teacher Education, 40*(4), 36–42.

Livingston, C. & Borko, H. (1990). High school mathematics review lessons: Expert/novice distinctions. *Journal for Research in Mathematics Education, 21*, 372–387.

National Council of Teachers of Mathematics (1989). *Curriculum and evaluation standards for school mathematics*. [Electronic Version] Reston, VA: Author.

National Council of Teachers of Mathematics (1991). *Professional standards for teaching mathematics*. [Electronic Version] Reston, VA: Author.

National Council of Teachers of Mathematics (1995). *Assessment standards for school mathematics*. [Electronic Version] Reston, VA: Author.

National Council of Teachers of Mathematics (2000). *Principles and standards for school mathematics*. Reston, VA: Author.

Putnam, R. & Borko, H. (2000). What do new views of knowledge and thinking have to say about research on teacher learning? *Educational Researcher, 21*(1), 4–15.

Schoenfeld, A. H. (1998). Towards a theory of teaching-in-context. *Issues in Education, 4*(1), 1–94.

Seago, N. (2004). Using video as an object of inquiry for mathematics teaching and learning. In J. Brophy (Ed.), *Advances in research on teaching: Vol. 10 Using video in teacher education* (pp. 235–286). Oxford, UK: Elsevier.

Shavelson, R. J. (1973). What is *the* basic teaching skill? *Journal of Teacher Education, 24*(2), 144–151.

Shavelson, R. J. (1976). Teachers' decision making. In N. L. Gage (Ed.), *The psychology of teaching methods*, Seventy-Fifth Yearbook of the National Society for the Study of Education, Part I. Chicago: University of Chicago Press.

Shavelson, R. J. (1986). Toma de decision interactive: algunas reflexiones sobre los procesos cognoscitivos de los profesores. (Translation: Interactive decision making: Some thoughts on teacher cognition.) In L. M. V. Angulo (Ed.), *Pensamientos de los profesores y toma de decisiones*. Universidad de Sevilla, Servicio de Publicaciones, Seville, Spain.

Shavelson, R. J. & Stern, P. (1981). Research on teachers' pedagogical thoughts, judgments, decisions, and behavior. *Review of Educational Research, 51*, 455–498.

Shavelson, R. J., Webb, N. M., & Burstein, L. (1986). Measurement of teaching. In M. C. Wittrock (Ed.), *Handbook of research on teaching* (3rd ed.). (pp. 50–91). Washington, DC: American Educational Research Association.

Sherin, M. G. (2004). New perspectives on the role of video in teacher education. In J. Brophy (Ed.), *Advances in research on teaching: Vol. 10 Using video in teacher education* (pp. 235–286). Oxford, UK: Elsevier.

Sherin, M. G. & Han, S. Y. (2004). Teaching learning in the context of a video club. *Teaching and Teacher Education, 20*, 163–183.

Shulman, L. S. (1986a). Paradigms and research programs in the study of teaching: A contemporary perspective. In M. C. Wittrock (Ed.), *Handbook of research on teaching* (3rd ed., (pp. 3–36). Washington, DC: American Educational Research Association.

Shulman, L. S. (1986b). Those who understand: Knowledge growth in teaching. *Educational Researcher, 15*(2), 4–14.

Shulman, L. S. & Elstein, A. S. (1975). Studies of problem solving, judgment, and decision making: Implications for educational research. In F. N. Kerlinger (Ed.), *Review of Research in Education* (Vol. 3). Itasca, IL: F.E. Peacock.

Stern, P. & Shavelson, R. J. (1983). Reading teachers' judgments, plans, and decision making. *The Reading Teacher, 37*, 280–286.

Wilson, S. M., Shulman, L. S., & Richert, A. E. (1987). "150 different ways" of knowing: Representations of knowledge in teaching. In J. Calderhead (Ed.), *Exploring teachers' thinking* (pp. 104–124). London: Cassell.

Section III
Spatial Abilities, Visualization, and Geometry

Bishop's work on spatial abilities started on his return from the USA in the mid 1960s. By then he had taken a position at the University of Hull and the initial work on this issue was carried out in conjunction with Frank Land. Bishop's research for his doctoral thesis, "Towards individualising mathematics teaching: An analysis and experimental investigation of some of the variables involved," overlapped with this area of research. He continued working on this issue through the 1970s after his move to Cambridge University.

The key article for this section was written as a summary article after some 15 years of thinking about how children's spatial abilities do impact on their mathematical learning, and the implications this interrelationship has for teaching. In it Bishop makes a number of references to his earlier work. A key notion for Bishop was that people did not have a singular spatial ability. This notion was plural: people have spatial abilit**ies**. In this article, Bishop also comments on research methodology, which he does often in his publications, reacting here in part to the still prevailing notions in education that the so-called objective number-based approach was the critical (and only?) research path to take. Bishop by now was wedded to a mixed methods research approach in which he does not dismiss factor analyses and such approaches out of hand, but argues that they are limited to specific questions, and other approaches are as important. From about this time, Bishop tended to move on to other areas of mathematics education research.

The two chapters in this section that reflect on Bishop's contribution to this aspect of research take different lines. Presmeg takes up the spatial abilities story where Bishop leaves it. She starts with her research for her doctorate, which Bishop supervised, and paints some of the broad avenues that have now opened up in the succeeding decades. Clements on the other hand focuses in on a particular time when Bishop was deciding to "move on" from his work in spatial ideas, after his experience with students in Papua New Guinea. It becomes a fascinating observation of one colleague by another: they are two colleagues who at this stage were only getting to know each other.

An Additional Bishop Reference Pertinent to This Issue

Bishop, A. J. (1983). Space and geometry. In R. Lesh & M. Landau (Eds.), *Acquisition of mathematics concepts and processes* (pp. 175–203). New York: Academic Press.

Reference

Land, F. W. (1964). *The language of mathematics.* New York: John Wiley.

Chapter 5
Spatial Abilities and Mathematics Education – A Review

Alan J. Bishop

Factor Analysis

The review begins with one of the main lines of research in the area—the work of the factor analysts. Historically, spatial abilities have been of interest ever since Galton (1883) began his systematic psychological inquiry. When, later on, the supposedly more objective methods of the factor analysts held supremacy over those of the introspectionists, researchers like Spearman (1927) and Thurstone (1938) attempted, by their increasingly more sophisticated statistical methods, to clarify in their various ways the structure of human intelligence. Their methods involved large scale group testing and the constructs were "ability" and "factor". Spatial ability and numerical ability were usually tested in all the large factor analysis studies but mathematical ability as such, was not. Spearman's view was that the solution of any task required the application of a general factor (general intelligence) and a specific factor. Mathematical ability seemed therefore to be conceived of as a combination of general intelligence applied to the mathematical context, and certain specific abilities such as numerical ability. Thurstone, on the other hand, did not hold with the notion of general factor but rather he proposed a set of Primary Mental Abilities which are required in certain combinations in any specific area like mathematical performance.

In the forties and fifties mathematical ability, or "the mathematical factor", became the focus of more studies, and researchers such as Barakat (1951), Werdelin (1958) and Wrigley (1958), wrestled with the problems of clarifying the nature of mathematical ability, and how it related to other abilities.

Unfortunately, their findings are unclear and certainly inconclusive. The relationships between spatial ability and mathematical ability differ from one study to another. MacFarlane Smith (1964), in his magnificent book, contributes a detailed analysis of these studies in one of his chapters and shows clearly his preference for a gestaltist view of spatial ability. He says for example, that the spatial loading of a test "depends on the degree to which it involves the perception, retention and recognition (or reproduction) of a figure or pattern in its correct proportions. Success in the item must depend *critically* on an ability to retain and recognise (or reproduce) a configuration as an organised whole" (p. 96). However, his case for arguing that

Educational Studies in Mathematics 11(1980) 257–269. 0013–1954/80/0113–0257$01.30

spatial ability is the key ability underlying mathematical ability has in this writer's opinion, still to be proved. For example in Werdelin's massive work, there were no significant loadings by the mathematics tests on the spatial factor.

It is of course, too easy to get drawn into the factor analysts' arguments over the uniqueness of the mathematical factor, or the correct combination of component abilities, or the hierarchical nature of abilities, or the complex methodological problems. The factor analysts, as their name suggests, are principally concerned with analysing intellectual factors and their relationships. They are part of the psychometric tradition which is based on the objective assessment and quantification of intellectual abilities. They make no reference to individuals, they rarely pursue how an individual approaches a solution to a particular problem, and they are rarely concerned with the classroom. Their legacy can be felt to be a disappointing one for mathematical educators, as Krutetskii (1976) clearly argues:

> It is hard to see how theory or practice can be enriched by, for instance, the research of Kennedy (1963), who computed, for 130 mathematically gifted adolescents, their scores on different kinds of tests and studied the correlation between them, finding that in some cases it was significant and in others not. The process of solution did not interest the investigator. But what rich material could be provided by a study of the processes of mathematical thinking in 130 mathematically able adolescents! (p. 14)

However, I would maintain that there is a great deal which has been, and can still be, gleaned from the work of the factor analysts. Firstly, they have produced many tests and these are available to other workers not just as tests, but also as examples of spatial tasks which could be used in individual testing, as elaborations and realizations of theoretical constructs, and as tasks for training spatial abilities. They can also stimulate the development of teaching material for classroom use.

Secondly, although the theoretical debates can appear arid and much removed from the curricular and classroom concerns of mathematics education, they can in fact help us in our own thinking. For example, Michael *et al.* (1957) showed by their analysis that the notion of 'spatial ability' as a unitary construct, with a gestaltist flavour, was inadequate for conceptualising the intellectual processes involved in the enormous range of 'spatial' tasks which had been developed. McGee (1979) presents a modified version of Michael's analysis, showing even more clearly the distinction between spatial visualization, where for example, the subject must imagine the rotations of objects in space, and spatial orientation, where the subject must recognise and comprehend the relationships between the various parts of a configuration and his own position.

This type of subdivision has been explored by others also. For example Guay and McDaniel (1977) investigated the relationship between high-level and low-level spatial abilities and elementary school mathematics achievement. They say:

> low-level spatial abilities were defined as requiring the visualization of two-dimensional configurations, but no mental transformations of those visual images. High-level spatial abilities were characterized as requiring the visualization of three-dimensional configurations, and the mental manipulation of these visual images. (p. 211)

Undoubtedly other analyses of spatial abilities are being, and will continue to be, developed (see for example Guay and Wattanawaha, 1978) but the notion of different spatial abilities (in the plural) remains. Rather like Krutetskii's use of "mathematical abilities", as opposed to the factor analysts' idea of 'mathematical ability' (in the singular), this change makes the construct (and its associated tasks) far more accessible for educational use. An 'ability' has the flavour of an individual difference, possibly inherited, but certainly given within each child. "Abilities", on the other hand, are described in more teachable terms, possibly capable of development within each child, but certainly there as objectives.

Thurstone's notions of Primary Mental Abilities offer us another provoking idea—if there are appropriate combinations of primary abilities which constitute mathematical ability, it could be that mathematical ability could be developed not by just teaching *mathematics*, but also by suitably emphasising and developing those primary abilities. This would have the effect of focussing our attention more on the abilities necessary for doing mathematics and away from the particular content of the mathematics curriculum.

Developmental Psychology

The last point suggests one way of approaching the development of mathematical abilities in children, and it is a point which has apparently escaped the attention of those who have seized upon the work of the developmental psychologists as being the only useful psychological tradition for the purposes of mathematical educators.

The massive work of Piaget and his followers is of course at the heart of this research tradition, particularly the books by Piaget and Inhelder (1956) and Piaget, Inhelder and Szeminska (1960). Moreover, despite experimental evidence which disputes some of the findings (see for example, Bryant and Trabasso, 1971), despite arguments about ages and stages (see for example, Donaldson, 1979), and despite detailed criticisms over the mathematical language used in the research (see for example, Martin, 1976b), the influence of the Geneva school is still formidable. Two recent publications on spatial and geometric concepts from the Georgia Center, Martin (1976a) and Lesh (1978), report research which was carried out almost entirely within the Piagetian tradition, with no reference being made to any of the factor analysts' work.

However, the Geneva school is not the only source for ideas regarding the development of spatial abilities in mathematics. The work of Werner (1964) is not as well known as it should be, though geography educators appear to recognise the importance of his notions of spatial development (see Hart and Moore, 1973). Within his *orthogenetic principle* there is a progression from a state of relative globality and lack of differentiation to a stage of increasing differentiation, articulation, and hierarchic integration. Werner's ideas clearly differ from Piaget's, though there are some similarities. In addition, Werner does not get into difficulties by using Piaget's 'mathematical' terms, which have to some measure both enabled mathematics educators to see a relevance for their subject, and also have obscured some of the real issues in Piaget's theory. Nor does one get the impression that he was quite so blinkered in his observations by his own theory. Finally, Werner's work reminds us that insofar as we are concerned with spatial ideas in mathematics as opposed to just visual ideas, we must attend to large, full-sized space, as well as to space as it is represented in models, and in drawings on paper. It is no coincidence that geography educators find his work of interest.

Another view of psychological development is that of Bruner (1964) whose proposed levels of representation "enactive, ikonic, symbolic" bear some relationship both to Piaget's stages and also to Werner's three levels "sensorimotor, perceptual and contemplative". Interestingly Bruner seemed to develop these ideas in the context of algebraic mathematics rather than in geometry. Also it occurs to me that were his first two levels to be fully exploited, there would be a similarity between this approach and the idea I took from the factor analysts' work of strengthening the 'component' abilities on which mathematical ability depends.

In contrast with the factor analysts then, there appears to be a discernible link between the interests of the developmental psychologist and the interests of the mathematics educator. Hence, despite protestations about the dangers of prescribing teaching methods and curricula based on Piagetian notions (see for example, Sullivan, 1967), we can see the evidence of the developmental psychologists' influence in elementary schools all over the world. It is likely of course that this is a benign influence, but one can also see the evidence of an unfavourable influence in those secondary schools where teachers make the assumption that all children are by that age at the formal operational stage. This is the problem, that there are also marked differences between the concerns of the developmental psychologist and the concerns of the mathematics educator. It must be pointed out that the former is essentially interested in the 'natural' development of the child, whereas the latter, because of his interventionist role, is essentially interested in 'unnatural' development.

Therefore, one can be guided by developmental psychologists' descriptions, but also one can ask why their theories should be accorded more importance than those devised by any other school of psychology? The tasks still remain for the mathematics educator of planning curricular sequences and developing teaching approaches which will achieve the goals of mathematics education. The work of van Hiele (1959) is an excellent example of the translations that need to be done and it is interesting that his descriptions seem to have more in common with Werner's ideas than with either Piaget's or Bruner's. Lesh (1976) also tackles this problem in the context of teaching transformational geometry in the elementary school.

Individual Differences

Of course the research methods of the two previous groups are very different, with the "clinical interview" being at the heart of the developmental psychologist's procedures. This is another reason why teachers feel some empathy with this school, one works with individual children. However there remains an essential disagreement – a developmental psychologist interprets the main differences between the children he interviews in terms of a different stage or level of development. An educator does not necessarily interpret the differences that way – indeed within any one class of children of a similar age, a teacher may not notice the differences of level, but he may well notice other differences. This points to a different school of research, the so-called "Individual Difference" tradition, which is also a long one, see Anastasi (1958).

Within the more specifically mathematics literature, the main differences which have been documented are the male/female differences (see Fennema, 1979), the gifted/less able differences (see Stanley *et al.*, 1974; Magne, 1979), the culture differences (see Lancy, 1979), and particularly relevant for our purposes here is that difference described by Krutetskii (1976) as the "analytic/geometric" difference.

This last individual difference – the extent to which an individual uses more, or less, visual ideas in solving mathematical problems – is one of the most frequently occurring differences in the mathematics education literature. The data range from accounts of problem solving by eminent mathematicians (see Hadamard, 1945), through various research reports such as Krutetskii's to teachers' descriptions of their own children's strategies (see Kent and Hedger, 1980). Krutetskii's contribution is an important one in at least three ways. Firstly, he has developed a set of tasks which include problems "involving a high degree of spatial thinking" and which make valuable connections between spatial abilities and mathematical abilities. Secondly, he documents several cases of pupils, good at mathematics, who use predominantly spatial ideas in their problem-solving. Thirdly, he shows us an example of a research method and style very different from both the psychometric and developmental traditions.

Typically, in individual difference research the first stage involves the documentation and description of the particular difference. The next stage is where, once again, we notice that the psychologist and the educator pursue different goals. The psychologist is usually concerned with why the differences exist, while the educator worries more about what should be done about them, if anything. So, there are many psychological studies involving within-individual reasons for individual differences, such as the person's heredity, their genetic composition, the lateral dominance in their brain, and hormonal and neurological factors. In the context of spatial ability differences, Harris' (1978) review is the most comprehensive available, although McGee's (1979) review does link these individual-difference studies with the factor analytic. At the present time mathematics educators do not appear to find a great deal there to develop within our own field though Wheatley and others (1978) have made some intriguing speculations concerning hemispheric specialisation and cognitive development.

It is when there is consideration of outside-individual variables as determinants of individual differences that the research moves more towards the educator's interest. Cultural contexts, environmental and social factors are all in the educator's domain, and all can offer ideas for the mathematics educator.

With regard to spatial ability differences there is a growing interest in cultural determinants and environmental factors. Berry's (1971) classic paper indicates which aspects of the learner's culture might affect the development of spatial skills – physical environment, language, occupational pursuits and social practices – but mathematical ability was not of concern in his study. Gay and Cole (1967) drew the attention of mathematics educators to the cultural constraints of learning, and more recently Bishop (1979) and Mitchelmore (1980) report research on the connections between mathematical and spatial abilities and the learner's culture. Mitchelmore's (1976) chapter offers an extensive review of the cross-cultural literature and points to many developments in testing and spatial training of interest not just to educators in developing countries.

One aspect of the learner's environment consists of the formal education he receives, and it is likely that teaching approaches are an important determinant of spatial abilities. Mitchelmore (1980) conjectures that differences in teaching approaches were responsible for the differences in 3D drawing ability which he found between West Indian, American and English children. As he says "English teachers tend to have a more informal approach to geometry, to use more manipulative materials in teaching arithmetic at the elementary level and to use diagrams more freely at both secondary and tertiary levels". (p. 8). Bishop (1973) found that children taught in primary schools where the use of manipulative materials predominated tended to perform better on spatial ability tests than children from 'material-free' primary schools. Fennema (1977) also pointed to the teaching when they suggested that the number of 'space-related' courses experienced by learners could have accounted for the differences between boys and girls on spatial tests. The boys tended to choose more space-related courses in school and this could have accounted for their higher spatial scores.

One complicating feature of individual difference research is that the researcher must accept conditions the way they exist. One cannot randomly assign subjects to cultures, to schools, or to social groups, and so the results of such research are never clear. However, the speculations which such differences provoke are important to the mathematics educator when considering not why the differences exist but what if anything should be done about them.

Teaching Experiments

Research which involves direct experimentation with teaching proceeds at present in two different directions. Both are concerned with what is possible to achieve but they are based on very different notions about what is desirable to teach.

The first approach seems to take, as its ideal, teaching which is individualised to the extent that all learners can make the maximum use of whatever abilities they possess. A major contribution to this approach has been made by research into aptitude-treatment-interactions (ATI). As its name implies this method explores those interactions which take place between different learner aptitudes and various (teaching) treatments. Radatz (1979) presents an interesting overview of this type of research, and two recent examples of ATI experiments are shown by Webb and Carry

(1975) and by Young and Becker (1979). It is a complex research technique with many methodological problems still to be solved, and the quality of the ideas which have emerged from the experiments are not yet commensurate with the quantity of work involved in carrying them out. Radatz feels that

> The often inconsistent results of studies of the relation of mathematics learning to individualized instruction, cognitive styles, specific abilities, or social environmental conditions suggest that at present a general instructional or mathematical learning theory is impossible. There can only be theories that apply to very specific conditions of mathematics instruction: the results are local descriptions or local theories: (pp. 361–362).

However, the definition and clarification of teaching methods which are being carried out in the course of preparing ATI experiments could well benefit the mathematics educator in the long term.

The other approach to individual differences in spatial abilities seems, to me at least, much more likely to yield fruitful ideas. This research is concerned with identifying, and then teaching, specific spatial abilities, and ranges from studies where the training takes the form of sophisticated coaching on problems very like the test items to controlled teaching experiments. An example of the former would be Dawson's (1967) work on training depth perception, which though successful used highly specific and non-generalizable methods. An example of the latter would be Brinkmann's (1966) research in which generalizable teaching methods were used with school children to produce gains on spatial test scores.

There are several reports of successes in the literature. Frandsen's (1969) research showed the positive effects of diagrammatic training with pupils of low spatial ability, and Saunderson (1973) achieved success also on spatial test scores, while Vladimirskii's (1971) work used a different approach in the context of geometry teaching. Marriott's (1978) work showed particularly interesting results of using a 'manipulative' kit for the teaching of ideas about fractions. The children who had used the kit, made errors on the final test which consisted of approximations based on their visualization of the teaching materials, e.g., $1/4 + 1/8 = 1/3$. Children in the control group, who did not use the kit, tended to make more computational errors such as $1/3 \times 3/5 = 14/15$. Thus, the manipulative materials, whilst not achieving any significant differences in total scores, nevertheless encouraged their users to visualize the problems and their solutions.

The methods of teaching used in this type of research come from various sources. Some, as in Brinkmann's study, are already in existence in school courses. Some are suggested by other work – for example, following the discovery of differences in children's spatial abilities as a result of their primary school teaching (Bishop, 1973), we used manipulative materials in a training programme with secondary school pupils, and achieved some quite high gains in their spatial task scores (Bishop, 1972). Other methods (such as in Dawson's study) come from detailed analyses of the sub-skills necessary for the successful completion of a spatial task.

Ideas such as these will probably be of no surprise to teachers in the technical education field, who have to teach technical drawing, for example. But it is my experience that the mathematics education community does not fully recognise the possibility or the desirability of such spatial training. However, a few people have recognised the need, and the promise of such research. Regarding girls' education, Sherman (1979) argues:

> Research needs to be directed towards factors affecting the development of spatial skill not only during early years, but even during adult years. We do not assume that an illiterate adult should be written off as unteachable, nor should we assume that adults cannot improve their spatial skill. Methods of achieving this . . . need to be devised, and their feasibility and advisability evaluated . . . For a swimmer with a weak kick we provide a kick board and opportunities to develop the legs. We do not further exercise the arms. (pp. 26–27)

Bruner (1973) also refers to the training of "subtle spatial imagery" and concludes "I don't think we have begun to scratch the surface of training in visualization". Mitchelmore (1976) in his review of cross-cultural research says "the greatest need is for the development of practical geometrical and spatial teaching programs and for their experimental testing". (p. 172).

Having, in a sense, shown that teaching spatial abilities is possible we need to move further. One therefore looks for developments which will generalise easily into the mathematics classroom context, for research into the transfer effects of such training, and most important of all, for more sensitive research which will uncover just which aspects of training programmes are responsible for their success. We have reports of no success in the literature, with both Ranucci (1952) and Brown (1954) finding that high school geometry courses didn't improve the learners' spatial test scores. Such findings are important. Perhaps successful teaching requires more detailed analysis resulting in a clear relationship between the teaching and the ability being taught. Clements (1978) paper presents a collection of spatial task analyses, which clearly offer great potential not only for research, but also for teaching.

Conclusion

I have tried to show that there are several different approaches to research at the mathematical/spatial interface and that it is a rich field for mathematics educators to study, provided that we are constantly aware of the dangers of becoming embedded in one research tradition. The factor analysts can offer tests and tasks, various classifications of abilities, and ideas about order and dependence between different abilities. The developmental psychologists can suggest ideas of what can be expected of children at different ages and of stages children will pass through (under present conditions) in learning to comprehend certain features of their world. The individual-difference researchers can offer many ideas about the reasons why these differences exist and how such differences could relate to the conditions in which learning and development takes place. Experiments in teaching can help to clarify just what these conditions are and how they can be exploited by teachers.

Finally, what have I omitted, and what issues remain? Certainly my grouping of the research into those four areas has led me to omit several sources for ideas. The most important are those of Shepard and Metzler (1971) on the mental rotation of three dimensional shapes, the field of "intuition" which has many connections with this review (see, for example, Fischbein *et al*, 1979), the work of Sheehan (1966) on vividness of imagery, and the research on information processing (see for example Neisser, 1970).

What issues remain? There are of course many, but perhaps I can select a few which seem significant to me:

1. How can one best determine and describe an individual's particular strengths and weaknesses? Specifically, how strong does an ability need to be, before it can be regarded as a strength? Conversely, how weak does it need to be before it is regarded as a weakness?
2. Do clinical observations of childrens' behaviour match classroom observations, or does the social nature of the classroom context significantly alter that behaviour? If it does significantly alter it what then is the value *for the classroom* of developmental theory based on clinical observation?
3. Should experimental teaching methods in this area take into account the spatial abilities of the teacher? If so, how?
4. I have shown examples of differences in 'levels' of spatial abilities, and I have indicated also the possible differences in teaching methods and training techniques. Do these two sets of differences relate together, i.e., do *training* techniques seem more appropriate for low-level 'skill'-type spatial abilities, and teaching methods seem more appropriate for high-level abilities (spatial cognition?)?
5. How much responsibility should mathematics teachers take for the training and teaching of spatial abilities? Is this perhaps an area like language, which is every teacher's responsibility? Is there perhaps a need for a core school course on 'graphicacy' (see Balchin, 1972)?

Department of Education
Cambridge University

References

Anastasi, A.: 1958, *Differential Psychology*, Macmillan, New York.

Balchin, W. G.: 1972, 'Graphicacy', *Geography,* 57, 185–195.

Barakat, M. K.: 1951, 'A factorial study of mathematical abilities', *British Journal of Psychology,* 4, 137–156.

Berry, J. W.: 1971, 'Ecological and cultural factors in spatial perceptual development', *Canadian Journal of Behavioural Science,* 3(4), 324–336.

Bishop, A. J.: 1972, 'Use of structural apparatus and spatial ability – a possible relationship', Transcribed talk given at the Psychology of Mathematics Education Workshop, Chelsea College, London.

Bishop, A. J.: 1973. 'The use of structural apparatus and spatial ability – a possible relationship', *Research in Education,* **9**, 43–49.

Bishop, A. J.: 1979, 'Visualising and mathematics in a pre-technological culture', *Educational Studies in Mathematics,* **10**, 135–146.

Brinkmann, E. H.: 1966, 'Programmed instruction as a technique for improving spatial visualization' *Journal of Applied Psychology* **50**, 179–184.

Brown, F. R.: 1954, *The Effect of an Experimental Course in Geometry on Ability to Visualize in Three Dimensions*. Unpublished doctoral dissertation, Columbia University: New York.

Bruner, J. S.: 1964, 'The course of cognitive growth', *American Psychologist* **19**, 1–15.

Bruner, J. S.: 1973, *Beyond the Information Given*, Allen and Unwin, London.

Bryant, P. E, and Trabasso, T.: 1971, 'Transitive inferences and memory in young children', *Nature*, **232**, 456–458.

Clements, M. A.: 1978, 'Some mathematics educators' view on spatial ability', in Low, B., Dunkley, M. E., and Conroy, J. (eds.), *Research in Mathematics Education in Australia, 1978*, Sydney, MacQuarie University.

Clements, M. A. and Wattanawaha, N: 1978, 'The classification of spatial tasks suitable for the classroom', in Williams, D.(ed.), *Learning and Applying Mathematics*, Australian Association of Mathematics Teachers, Melbourne.

Dawson, J. L. M.: 1967, 'Cultural and physiological influences upon spatial-perceptual processes in West Africa. Part 1', *International Journal of Psychology*, **2**, 115–128.

Donaldson, M.: 1978, *Children's Minds*, Fontana/Collins, Glasgow.

Fennema, E.: 1979, 'Women and girls in mathematics – equity in mathematics education', *Educational Studies in Mathematics*, **10**, 389–401.

Fennema, E. and Sherman, J. A.: 1977, 'Sex-related differences in mathematics achievement, spatial visualization and socio-cultural factors', *American Educational Research Journal* **14**, 51–71.

Fischbein, E., Tirosh, D., and Hess, P.: 1979, 'The intuition of infinity', *Educational Studies in Mathematics*, **10**, 3–40.

Frandsen, A. N. and Holder, J. R.: 1969, 'Spatial visualization in solving complex verbal problems', *Journal of Psychology*, 73, 229–233.

Galton, F.: 1883, *Inquiries into the Human Faculty and its Development*, Macmillan, London.

Gay, J. and Cole, M.: 1967, *New Mathematics in an Old Culture*, Holt, Rinehart and Winston, New York.

Guay, R. B. and McDaniel, E. D.: 1977. 'The relationship between mathematics achievement and spatial abilities among elementary school children', *Journal for Research in Mathematics Education* 8, 211–215.

Hadamard, J.: 1945, *The Psychology of Invention in the Mathematical Field*, Dover, London.

Harris, L. T.: 1978, 'Sex differences in spatial ability: Possible environmental, genetic and neurological factors', in Kinsbourne, M. (ed.), *Assymetrical Functions of the Brain*, Cambridge University Press, Cambridge, England.

Hart, R. A. and Moore, G. T.: 1973, 'The development of spatial cognition: a review', in Downs, R. M. and Stea, D.(eds,), *Image and Environment: Cognitive Mapping and Spatial Behaviour*, Arnold, London.

van Hiele, P. M.: 1959, 'La pensée de l'enfant et la géometrie', *Bulletin de l'Association des Professeurs de l'Enseignement Public* **198**, 199–205.

Kennedy, W. A., Willcutt, H. and Smith, A.: 1963, 'Wechsler profiles of mathematically gifted adolescents'. *Psychological Reports* **12**, 259–262.

Kent, D. and Hedger, K.: 1980, 'Growing tall', *Educational Studies in Mathematics* **11**, 137–179.

Krutetskii, V. A.: 1976, *The Psychology of Mathematical Abilities in School Children*, University of Chicago Press, Chicago.

Lancy, D. F. (ed.): 1978, *The Indigenous Mathematics Project*, special issue of the *Papua New Guinea Journal of Education* **14**.

Lesh, R.: 1976, 'Transformation geometry in elementary school: some research issues' in Martin (1976a), pp. 185–243.

Lesh, R. (ed.): 1978, *Recent Research Concerning the Development of Spatial and Geometric Concepts*, ERIC/SMEAC, Columbus, Ohio.

MacFarlane Smith, I.: 1964, *Spatial Ability: Its Educational and Social Significance*, University of London Press, London.

Magne, O.: 1978, 'The psychology of remedial mathematics', *Didakometry* **59**, 1–27.

Marriott, P.: 1978, 'Fractions: now you see them, now you don't', in Williams, D. (ed.), *Learning and Applying Mathematics,* Australian Association of Mathematics Teachers, Melbourne.

Martin, J. L. (ed.): 1976a, *Space and Geometry,* ERIC/SMEAC, Columbus, Ohio.

Martin, J. L.: 1976b, 'An analysis of some of Piaget's topological tasks from a mathematical point of view', *Journal for Research in Mathematics Education,* **7**, 8–24.

McGee, M. G.: (1979). 'Human spatial abilities: Psychometric studies and environmental, genetic, hormonal, and neurological influences', *Psychological Bulletin,* **86**, 889–918.

Michael, W. B., Guilford, J. P., Fruchter, B., and Zimmerman, W. S.: 1957, 'The description of spatial-visualization abilities', *Educational and Psychological Measurement,* **17**, 185–199.

Mitchelmore, M. C.:1976, 'Cross-cultural research on concepts of space and geometry', in Martin (1976a), pp. 143–184.

Mitchelmore, M. C.: 1980, 'Three dimensional geometrical drawing in three cultures', *Educational Studies in Mathematics* **11**, 205–216.

Neisser, U.:1970, 'Visual image as process and as experience', in . Antrobus, J.S.(ed.), *Cognition and Effect,* Little, Brown, Boston.

Piaget, J. and Inhelder, B.: 1956, *The Child's Conception of Space,* Routledge and Kegan Paul, London.

Piaget, J., Inhelder, B. and Szeminska, A.: 1960, *The Child's Conception of Geometry,* Routledge and Kegan Paul, London.

Radatz, H.: 1979, 'Some aspects of individual differences in mathematics instruction', *Journal for Research in Mathematics Education* **10**, 359–363.

Ranucci, E. R.: 1952, *The Effect of the Study of Solid Geometry on Certain Aspects of Space Perception Abilities*. Unpublished doctoral dissertation, Columbia University, New York.

Saunderson, A.: 1973, 'The effect of a special training programme on spatial ability test performance', *New Guinea Psychologist 5*, 15–23.

Sheehan, P. W.: (1966), 'Functional similarity of imaging to perceiving: Individual differences in vividness of imagery', *Perceptual and Motor Skills* **23**, 1011–1033.

Shepard, R. N. and Metzler, J.: 1971, 'Mental rotation of three-dimensional objects', *Science* **171**, 701–703.

Sherman, J. A.: (1979), *Women and Mathematics: Summary of Research from 1977–1979 NIE Grant*. Unpublished manuscript.

Spearman, C. E.: 1927, *The Abilities of Man: Their Nature and Measurement*, Macmillan, London.

Stanley, J. C., Keating, D. P. and Fox, L. H.: 1974, *Mathematical Talent: Discovery, Description and Development,* John's Hopkins, University Press, Baltimore.

Sullivan, E. V.: 1967, *Piaget and the School Curriculum: A Critical Appraisal*, Bulletin 2, Ontario Institute for Studies in Education.

Thurstone, L. L.: 1938, Primary Mental Abilities, *Psychometric Monographs*, 1–121.

Vladimirskii, G. A.: 1971, 'An experimental verification of a method and system of exercises for developing spatial imagination', in Kilpatrick, J. and Wirszup, I. (eds.), *Soviet Studies in the Psychology of Learning and Teaching Mathematics 5*, Vroman, California.

Webb, L. F. and Carry, L. R.: 1975, 'Interaction of spatial visualization and general reasoning abilities with instructional treatment in quadratic inequalities: A follow-up study', *Journal for Research in Mathematics Education* **6**, 132–140.

Werdelin, I.: 1958, *The Mathematical Ability: Experimental and Factorial Studies*, Gleerups, Lund.

Werner, H.: 1964, *Comparative Psychology of Mental Development*, International Univerisities Press, New York.

Wheatley, G. H. Mitcheli, R., Frankland, R. L. and Kraft, R.: 1978, 'Hemispheric specialization and cognitive development: Implications for mathematics education', *Journal for Research in Mathematics Education* **9**, 19–32.

Wrigley, J.: 1958, 'The factorial nature of ability in elementary mathematics', *British Journal of Educational Psychology* **28**, 61–78.

Young, C.D and Becker, J.P.: 1979, 'The interaction of cobnitive aptitudes with sequences of figural and symbolic treatments of mathematical inequalities', *Journal for Research in Mathematics Education* **10**, 24–36.

Chapter 6
Spatial Abilities Research as a Foundation for Visualization in Teaching and Learning Mathematics

Norma Presmeg

Alan Bishop's review of psychological research on spatial abilities in 1980 presented the state-of-the-art in the field at that time. However, it did much more than that. In line with his lifelong concern for improving the teaching and learning of mathematics, he identified salient elements and threads from these studies that would be developed and investigated in research in mathematics education for years to come. In this sense, his 1980 paper in *Educational Studies in Mathematics* provided a foundation for research on visualization in mathematics education that continued through the decades of the 1980s and 1990s and is ongoing today, as witnessed by a plenary paper in the proceedings of the 30th conference of the International Group for the Psychology of Mathematics Education (Presmeg, 2006b), with the title, "A semiotic view of the role of imagery and inscriptions in mathematics teaching and learning."

Inevitably, writing about Alan's work will implicate my own. He was an inspiring supervisor for my doctoral research at Cambridge University, 1982–1985. Since then, his influence has been apparent in many aspects of my research, both in the area of visualization in mathematics education, and in the role of culture in teaching and learning mathematics. However, it is the former, his work and influence on visualization research, that is the topic of this chapter. The threads that form the backbone of this chapter celebrating Bishop's work on spatial abilities, visualization, and the teaching and learning of geometry, are as follows.

A Century of Psychological Research on Spatial Abilities

Already in 1980, Alan Bishop was recognizing the need to move beyond the field of psychological research in ascertaining the various elements that influence individual learning of mathematics, including those related to spatial abilities (with a stress on the plural form rather than the singular, ability). Despite the psychological

N. Presmeg
Illinois State University, Department of Mathematics, 313 Stevenson Hall, Normal IL 61790-4520, USA
e-mail: npresmeg@msn.com

P. Clarkson, N. Presmeg (eds.), *Critical Issues in Mathematics Education*,

focus of his 1980 paper – going right back to Galton's work in the late 1800s – his mention of social and cultural issues (Bishop, 1980, p. 262 & especially p. 266) foreshadows his later writings that greatly developed these ideas and contributed to a broadened theoretical lens that included sociocultural and political elements and the role of values in research on the teaching and learning of mathematics (see later chapters in this volume). As Clements elaborates in his chapter, Bishop had already worked with three culturally diverse groups of students in papua New Guinea, and his thinking had been influenced by this experience, as evidenced by his descriptions of this research (e.g., Bishop, 1983; Clements, chapter 7, this volume). It is not an exaggeration to state that his seminal thinking in these sociocultural areas, evident already in 1980 and growing through the decade of the 1980s (Bishop, 1988a, 1988b) helped to establish a new paradigm that moved away from major reliance on psychology in mathematics education. However, researchers in mathematics education did not leave psychology completely. The individual is still an important focus in this research, and research on visualization in teaching and learning mathematics has continued in a psychological format although the potentially useful theoretical lenses available have broadened considerably (Presmeg, 2006a).

It is the psychological emphases in his four subheadings, Factor Analysis, Developmental Psychology, Individual Differences, and Teaching Experiments that are interrogated and expanded in this chapter from the viewpoint of current research on visualization and geometry teaching and learning. In 1980, factor analysis and research in developmental psychology were the overarching paradigms in psychological investigations. Bishop juxtaposed and contrasted the methods of these two schools of thought, revealing awareness of the strengths and limitations of both while still seeing potential relevance of each for research on spatial abilities in mathematics education, where as yet there were few investigations in this area. Individual differences of various kinds were already quite extensively researched in the field of psychology. Gender differences, and variations in individual cognitive processing, were two areas that were seen to have implications for mathematics education, and in this young field (only the fourth International Congress on Mathematical Education, held every 4 years, took place in 1980 at Berkeley, California), the harvest was ready. However, it was the fourth subheading, concerning teaching, that was always close to Bishop's heart. He rightly pointed out that there are differences in the concerns of psychologists and mathematics educators. With regard to individual variations in cognitive processing, he explained that "The psychologist is usually concerned with why the differences exist, while the educator worries more about what should be done about, them, if anything" (Bishop, 1980, p. 262). There is an interventionist aspect to a mathematics educator's work, which was not paramount in reported psychological research at the time. It will be seen throughout this chapter that dedication to the improvement of the learning of mathematics through careful, rigorous research into aspects of teaching and learning, was at the heart of Bishop's work. His writing, in an easy, almost colloquial style, was the carrier of a strong sense of structure and organization, and the 1980 paper in Educational Studies in Mathematics is a good example of these characteristics.

The Problematic Construct of "Transfer of Training"

In the early 1980s, writers such as Bishop did not have the benefit of later insights that cast doubt on the power of learning to transfer to new contexts. Writing of Thurstone's Primary Mental Abilities, Bishop (1980, p. 259) suggested that mathematical learning could be enhanced by "suitably emphasising and developing those primary abilities," which include spatial abilities. He repeated the suggestion in an even stronger form later in the paper (p. 265), citing a collection of spatial tasks developed by Ken Clements (1978) that held promise for detailed analysis that could yield positive transfer, despite other research studies in which the results were negative. Careful attention to relevant detail in the tasks seemed to be the key for transfer to occur. The problem of continuity of cognition across settings is crucially important for education (Kirshner & Whitson, 1997). In recent decades, mathematics education researchers are far more cognizant of the situated nature of learning, informed by studies such as those of Lave (1997). Whether or not there is direct transfer of "training" (skills) or of "education" (learning dispositions), Bishop's writing reflects his constant concern that research in mathematics education should be relevant to the nitty-gritty details of the teacher's work in the mathematics classroom.

An extension of Bishop's (1980) concern to create finer distinctions in analyzing both the tasks used and the results of research on spatial aspects of mathematics education is evident in his proposal of two types of ability constructs, namely, interpreting figural information (IFI), and visual processing (VP). He elaborated on these constructs and their use in a later paper on Space and Geometry (Bishop, 1983). IFI refers to "understanding the visual representations and spatial vocabulary used in geometric work, graphs, charts, and diagrams of all types," whereas the more dynamic VP "involves visualization and the translation of abstract relationships and nonfigural information into visual terms" as well as "the manipulation and transformation of visual representations and visual imagery" (p. 184). IFI is an ability that relies on understanding of content and context, which relates particularly to the form of the stimulus material. In contrast, VP does not relate to the form of the stimulus material because it is an ability of process rather than content. These distinctions proved useful when Lean (1981) summarized the literature on "spatial training." Characteristically, Bishop (1983) asked two questions regarding VP, as follows.

1. Is it teachable?
2. If it is done within geometry, does it transfer to arithmetic and algebra?

Krutetskii's (1976) research was relevant to the first of these questions, suggesting that the answer is affirmative to some extent, although individual preferences remain. Lean and Clements (1981) took the question further, suggesting that figural and nonfigural stimuli should be used in teaching that encourages VP. The question of transfer was left open.

Although these constructs did not turn out to be central in my own research (Presmeg, 1985), the careful analytical distinctions advocated by my dissertation supervisor, Alan Bishop, comprised an exemplary model that I attempted to follow in taking his ideas further and adapting them to my investigation. Working with 13 high school mathematics teachers and 54 "visualizers" in their classes over a complete school year suggested that despite mnemonic advantages, unresolved problems with the generalization of visual and spatial information could hamper students' learning of mathematics in all content areas. These issues went beyond questions of transfer. The teaching significance of this research is pursued in a later section of this chapter.

Research on "Large, Full-Sized Space"

In the section on developmental psychology in Bishop's article, in comparison and contrast with Piagetian theory, Bishop (1980, p. 260) referred to Werner's (1964) *orthogenetic principle*, "a progression from a state of relative globality and lack of differentiation to a stage of increasing differentiation, articulation, and hierarchic integration." Werner worked in the content field of geography, in which a focus on "large, full-sized space, as well as . . . space as it is represented in models, and in drawings on paper" (Bishop, p. 260) is natural. Bishop reminded us that these foci are no less important in the field of mathematics education, and should not be neglected. That his admonition has been heeded in the years since then is evident in the chapters of a volume on *Symbolizing, modeling and tool use in mathematics education* (Gravemeijer, Lehrer, van Oers, & Verschaffel, 2002). The chapter by Lehrer and Pritchard (2002) is of particular significance in their description of third-grade children modeling the large-scale space of their playground using mathematical ideas of scale, origin, and coordinates to describe position and direction.

Philip Clarkson reminisces that he was starting to use photographs of outside scenes in his teaching in the 1970s, as a stimulus for teachers to use with their students in mathematizing the environment. It seemed to him "just a good teaching thing to do", but when he teamed up with Ken Clements and Alan visited Melbourne, the practice was reinforced and given a strong theoretical basis. More recently, pictures from the web are freely available, or students might be given cameras and asked to take their own pictures of scenes in which they identify mathematics.

Bruner's "Enactive, Ikonic and Symbolic" Modes, and Other Theories

Referring to Bruner's (1964) characterization of these three modes of representation, Bishop (1980) compared them with the developmental hierarchy of Piaget's stages, as well as with Werner's "sensorimotor, perceptual, and contemplative" levels (p. 260). In order to investigate whether current research in mathematics education had established that Bruner's enactive, iconic, and symbolic modes of representation did indeed satisfy a developmental hierarchy, Presmeg (2006a) examined relevant research studies over the last three decades. In no study was this conjecture verified. She concluded as follows.

> From the foregoing, it seems clear that individual differences in types of imagery, quality and quantity, preference for and skill in using, persist through the school years and possibly through lifetimes, without evidence of general developmental trends in forms of imagery or in their personal use. Bruner's (1964) well known *enactive, iconic*, and *symbolic* modes of cognition should therefore be taken as metaphors for types of thinking rather than as a developmental hierarchy.
>
> (p. 223)

With his sensitivity to the pitfalls of accepting and using stages or levels blindly when working with children, Bishop would probably endorse this conclusion.

Bishop (1980) also commented on the "massive work of Piaget", but at that time the neglect in mathematics education of some theories that are equally promising for the fostering of visual and spatial skills, such as the work of Werner and van Hiele. Van Hiele's framework for levels of thinking in geometry – which may be characterized as recognition, analysis, ordering, deduction, and rigor – has since then received quite considerable research attention (e.g., Gutierrez, Jaime, & Fortuny, 1991). However, the potential of the *phases* van Hiele outlined between each level and the next, to encourage individuals' growth in geometric thinking, has as yet not been fully exploited. Again referring to the preponderance of research on geometry that uses a Piagetian framework, in his later chapter Bishop (1983) commented on the sequence of topological, projective, and Euclidean thinking of children learning geometry posited by Piaget. As a hierarchy, he pointed out that this sequence and the mathematical terminology used by Piaget in this regard have proved problematic, and that excessive attention to frameworks such as this one may serve to obscure frameworks that might prove to be more productive. It is noteworthy that since then geometry researchers, particularly in Italy, have adopted other lenses such as those provided by Vygotskian thought (Mariotti, 2002).

Krutetskii's Powerful Influence in Research on Individual Differences

As Bishop (1980, pp. 261–262) pointed out, research on individual differences had a long tradition in psychology. In mathematics education, gender differences and differences in preferred styles had been investigated, as well as cognitive differences according to abilities. In particular, the translation of the writings of Krutetskii (1969, 1976) from their original Russian gave promise of a large influence on research into individual differences in the learning of mathematics. The importance of Krutetskii's research lay in its distinction between *level* of mathematical abilities, determined largely by a verbal-logical component of thinking, and *type* of mathematical cognition, determined largely by a visual-pictorial component. In the case of the latter, it is not only the ability to use visual-spatial thinking, but preference for its use, which determines the type of mathematical cognition that individuals employ in learning and doing mathematics. Rather than contrasting visual and analytical thinking in a continuum as others had done (e.g., Lean & Clements, 1981), Krutetskii's model posited two orthogonal axes, one for logical or

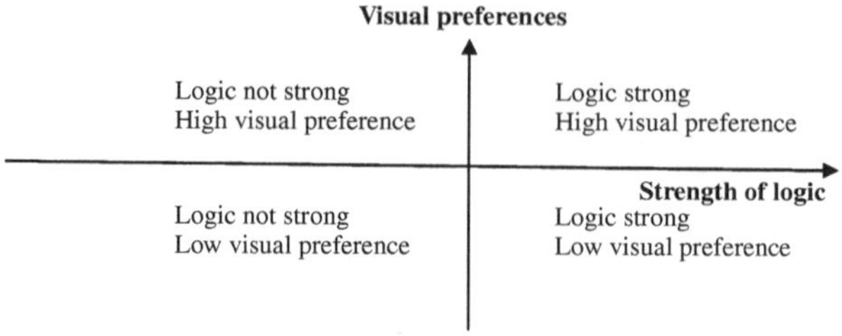

Fig. 6.1 Quadrants formed by separating logical and visual components of mathematical abilities

analytic thinking, and the other for the amount of visual thinking used by an individual, in mathematical problem solving. Spatial ability is not sufficient to ensure that an individual may *prefer* to solve mathematical problems using visual means. The logical or rational component is the defining factor in mathematical success. My research (Presmeg, 1985) confirmed that there were individuals at high school level whose mathematical thinking could be placed in all four of the quadrants so formed (Fig. 6.1), providing further substantiation for Krutetskii's model.

Bishop (1980) pointed out that Krutetskii's contribution was also significant for his substantial collection of mathematical interview tasks, for his carefully described case studies of the thinking of "capable" mathematics students, and for "a research style very different from both the psychometric and developmental traditions" (p. 262). Krutetskii's collection of mathematical tasks, his "system of experimental problems" (Krutetskii, 1976, p. 100) comprises 26 series of problems grouped in the following categories.

Information gathering:	Perception (interpretation of a problem) – 4 series.
Information processing:	Generalization – 8 series.
	Flexibility of thinking – 4 series.
	Reversibility of mental process – 1 series.
	Understanding; reasoning; logic – 4 series.
Information retention:	Mathematical memory – 1 series.
Typology:	Types of mathematical ability – 4 series.

This system of carefully categorized mathematical problems (77 pages of them in the 26 series, Chapter 8, pp. 98–174) is still a treasure trove for researchers in mathematics education, a rich resource that has not been fully exploited. Changes in research styles are elaborated in the next section.

Changes in Research Methodologies

Through the decade of the 1980s, the still dominant psychometric tradition of research in the Western hemisphere was gradually giving way, in mathematics education as in related fields such as science education, to a more humanistic tradition that employed qualitative methodologies. Aptitude-Treatment-Interaction studies (Bishop, 1980, p. 264), and formal teaching experiments that involved hypothesis testing based on statistical analyses, had been the scientific standard for rigorous research. In a powerful critique of such methods, Krutetskii (1976, p. 14) had argued as follows:

> It is hard to see how theory or practice can be enriched by, for instance, the research of Kennedy, who computed, for 130 mathematically gifted adolescents, their scores on different kinds of tasks and studies the correlation between them, finding that in some cases it was significant and in others not. The process of solution did not interest the investigator. But what rich material could be provided by a study of the process of mathematical thinking in 130 mathematically able adolescents!
>
> (Cited by Bishop, 1980, p. 258)

Krutetskii's methodology was a "think-aloud" interview procedure, followed by analysis and carefully documented case studies of the thinking of individual students, whose processing was characteristic of categories that he identified. Although Bishop (1980) cited Krutetskii's objections approvingly, with characteristic balance he also suggested that there was still room for the kind of research that involves quantification and testing, of course with a different goal in mind for the investigation. However, in the ensuing two decades, in a vast change of conceptions about the value of such research for the work of teachers, methodologies swung to case studies involving clinical interviewing and observation, and ethnographic studies after the mode of anthropological research, although not all mathematics education studies calling themselves ethnographic embraced the full rigor of anthropological investigation (Denzin & Lincoln, 2000). Krutetskii's (1976) task-based clinical interviews, in which individuals were asked to "think aloud" as they solved mathematical problems, provided a strong methodology for the new paradigm.

Using Krutetskii's (1976) model as one of three theoretical lenses, my research in the 1980s (Presmeg, 1985) used both quantitative and qualitative methods. Nonparametric statistical methods were useful in designing a test for preference for visualization in mathematics and interpreting its results. However, the depth of understanding of the strengths and pitfalls of visual thinking in individual high school mathematics students, alone and in interaction with their teachers, came from the qualitative interviews with students and teachers. In the 1990s the pendulum swung far in favour of qualitative research.

Now in the first decade of the 2000s there is a more balanced perception that these contrasting methodologies have different purposes, and that mixed methods can in fact draw on the strengths of both. In recent years it has become acceptable in mathematics education research to use a methodology of mixed methods (Johnson & Onwuegbuzie, 2004), in which the scientific rigor of statistical research is perceived as complementary to the intuitive insights that are possible in fine-tuned qualitative research. Each addresses different questions, and serves different functions. In mixed-methods research, going beyond the significance for different stakeholders that Cobb (2007) identified, an investigation may address the details of some educational phenomenon and attempt to generalize by identifying, for instance, how widespread the phenomenon is. Johnson and Onwuegbuzie presented an eight-step process for conducting such research, which they consider superior to mono-method research. As more mixed-method investigations appear in mathematics education research it will be interesting to see whether they have significance for both of the groups identified by Cobb (2007) – policymakers and administrators (the audience he identified for statistical research) as well as classroom teachers of mathematics (the audience for qualitative and mildly quantitative investigations). What counts as "good" educational research? Hostetler (2005) encouraged researchers to move beyond questions of qualitative and quantitative paradigms, and to consider the ethical and moral values entailed in research methodologies – a position that Bishop would be likely to endorse. The values entailed in mathematics education, and also in various research methodologies, were aspects that were developing in Bishop's thought from the 1980s onwards, as other chapters in this volume elaborate.

Bishop recognized the power of using mixed methods research in mathematics education already in 1980.

The Constant Teaching Thread

I have already mentioned Bishop's concern that mathematics education research should be relevant and useful to teachers in their daily work. He perceived (1980, p. 261) an "essential disagreement" between the concern of a developmental psychologist to identify the stage or level in which a student might be placed, and the concern of a mathematics educator to foster a learner's mathematical growth, that is, to *change* the level on which a child might be thinking. One is reminded of a new kind of teaching experiment that developed in the 1990s, based on the theoretical foundation of radical constructivism, in which one-on-one interviews with children are conducted and carefully video-recorded, for the purpose of documenting the changes that happen when the *purpose* of the interviews is learning, i.e., change, rather than documenting the status quo (Steffe & Thompson, 2000). Multitiered teaching experiments and design research (Kelly & Lesh, 2000) reflect a current concern to work with teachers in research that addresses mathematics curricula as well as pedagogy.

During the years 1982–1985 when Alan Bishop was supervising my research at Cambridge University, he expressed the opinion several times that the aspect of my research that dealt with teaching and classroom aspects facilitative of visual thinking was the most significant part. Although the visual thinking of individual students has received more attention (e.g., their use of *pattern imagery* and of metaphors in order to attain mathematical generalization – Presmeg, 1992, 1997), it is still the teaching aspects that serve to fill a lacuna in the literature (Presmeg, 2006a). Researchers such as Owens (1999) and Gray (1999) have taken up the issue of *teaching* spatial visualization and developing curriculum materials for young children. Their research results were reported in 1999 in a Research Forum on *Visual thinking in mathematics education*, at the 23rd Annual Meeting of the International

Group for the Psychology of Mathematics Education (PME). There is room for more development in this field.

The changes enabled by dynamic computer environments (e.g., Parzysz, 1999; Yerushalmy, Shternberg, and Gilead, 1999; at the same PME Research Forum) could provide a chapter in their own right: they will merely be mentioned here. The work of Parzysz on spatial visualization in geometry continued through the decade of the 1990s, providing illuminating insights on dynamic visualization in high school geometry. Yerushalmy and her collaborators provided a fine-toothed analysis of types of algebra problems, and how the use of dynamic computer software could deepen the understanding of researchers interested in students' learning of algebra. That Bishop (1983) saw the potential of technology to aid in overcoming the "specificity of a diagram" (p. 180) before dynamic computer software existed, is attested by his statement that "It would seem valuable to use film to clarify the meaning of some generalized geometric relationships" (ibid.).

The Issues that were Omitted in the 1980 Review

Some of the topics that were addressed in psychological literature at the time had not yet been investigated in relation to their bearing on spatial abilities in mathematics education. These included "the mental rotation of three dimensional shapes" (Shepard & Metzler, 1971, quoted in Bishop, 1980, p. 266), and Neisser's (1970) research on vividness of imagery (ibid.). Both of these aspects of visualization were investigated in mathematics education in the intervening decades. Mental rotation of shapes was an integral part of Wheatley's (1997) test for spatial ability, and of his research using an interview methodology with elementary school learners.

Vividness of imagery was a contributing factor to the mnemonic advantages of visual imagery, although it was not essential, in the cognition of the high school students in Presmeg's (1985, 1997) research. Bishop (1980) did not mention the controllability of mental imagery (Richardson, 1969, 1972). However, this aspect also turned out to be important in avoiding pitfalls of visualization in Presmeg's (1985) high school study. Further, controllability was an issue in the visual thinking of a student in a calculus class at the collegiate level (Aspinwall, Shaw, & Presmeg, 1997). In Aspinwall's dissertation research, university student Tim (pseudonym) was convinced by an uncontrollably persistent image that the graph of a quadratic function had to have a vertical asymptote: thus he struggled to draw the graph of the derivative of this function. Aspinwall's research with calculus students at the university level is ongoing.

Issues that Remain: The Significance of Bishop's Work

With regard to the genesis and evolution of thought in mathematics education, some of the constructs of Charles Sanders Peirce are relevant. One of the tenets in Peirce's epistemology is the continuity of past, present, and future. Continuity is central in Peirce's definition of synechism as "the tendency to regard continuity . . . as an idea of prime importance in philosophy" (Peirce, 1992, p. 313). Synechism involves the startling notion that knowledge in its real essence depends on future thought and how it will evolve in the community of thinkers:

> Finally, as what anything really is, is what it may finally be come to be known to be in the ideal state of complete information, so that reality depends on the ultimate decision of the community; so thought is what it is, only by virtue of its addressing a future thought which is in its value as thought identical with it, though more developed. In this way, the existence of thought now, depends on what is to be hereafter; so that it has only a potential existence, dependent on the future thought of the community.
>
> (Peirce, 1992, pp. 54–55)

Whether "the ideal state of complete information" is ever an attainable goal, is a matter of doubt, but the relevance of synechism for the history of mathematics education lies in the role attributed to future generations of thinkers in assessing the achievements of the past and present. Peirce (1992) cast further light on what he meant by continuity in his law of mind:

> Logical analysis applied to mental phenomena shows that there is but one law of mind, namely, that ideas tend to spread continuously and to affect certain others which stand to them in a peculiar relation of affectability. In this spreading they lose intensity, and especially the power of affecting others, but gain generality and become welded with other ideas.
>
> (p. 313)

Because of the importance of personal interpretations in forging a community of thinkers with its conventions, and thus in the continuity of ideas, Peirce formulated three kinds of interpretant in his semiotic model, which he designated, intensional, effectual, and communicational. The communicational interpretant, or commens, is "a determination of that mind into which the minds of utterer and interpreter have to be fused in order that any communication should take place" (Peirce, 1998, p. 478). For the continuity of ideas and their evolution, a central requirement is that there be a community of thinkers who share a "fused mind" sufficiently to communicate effectively with one another – and with posterity through their artifacts – through this commens.

In asking to what extent a community of fused minds has formed around the commens of Bishop's (1980) ideas on spatial abilities and mathematics education, it is informative to examine relevant chapters in some handbooks on research in mathematics education. In the early and influential Handbook of research on mathematics teaching and learning (Grouws, 1992), the chapter on "Geometry and spatial learning" by Clements and Battista cited both of Bishop's relevant earlier publications (1980 and 1983) as well as his "Review of research on visualization in mathematics education" (Bishop, 1989). It is noteworthy that in the International handbook of mathematics education (Bishop, Clements, Keitel, Kilpatrick, & Laborde, 1996), his work is not cited in the chapter on "Space and shape," although it is well represented in later chapters on culture, language, and the anthropology, sociology, and politics of mathematics education. Bishop had moved on to the other areas in which his thought has been influential. It appears that Peirce's (1992) law of mind was at work, namely, the principle that ideas tend to spread continuously and to affect certain others, until they lose intensity, and especially the power of affecting others, but gain generality and become welded with other ideas. Bishop's early work on spatial abilities and geometry is also not mentioned in the relevant chapter in the Second handbook of research on mathematics teaching and learning (Lester, 2007), and it is difficult to ascertain in the Handbook of international research in mathematics education (English, 2002) because the organization in this handbook was not accomplished through content classification. Despite these omissions, I would argue that a community sharing a commens did develop around Bishop's spatial publications. Although these ideas may seem to have lost their original intensity, they have evolved to such a degree that the original is no longer easily discernable. The evidence for this claim is provided in the following paragraph.

In the intervening decades, the concern to investigate visual aspects of teaching and learning mathematics has not diminished (Presmeg, 2006a). However, the focus has widened to embrace representation in more general terms (Goldin & Janvier, 1998; Hitt, 2002). In this broadening, the association of imagery with affect has received more systematic analysis and theoretical attention, although its significance was already noted in the early research on visualization in the 1980s (Presmeg, 1985). The theoretical lenses used to design and interpret the results of research on mathematical representation have also multiplied. One such significant lens is semiotics, which is currently receiving increased attention in mathematics education research (Anderson, Sáenz-Ludlow, Zellweger, & Cifarelli, (2003). In his "Guidelines for future research," Bishop (1983) suggested two areas that have been well addressed and have become almost commonplace, namely, real life "activities that scale down the environment" and "features such as the extent of active manipulation necessary, and the value of discussion with other children" (p. 198). The former is well represented in research carried out at the Freudenthal Institute (e.g., Gravemeijer et al., 2002), and the latter is so broadly investigated now that it is enshrined in the curricular recommendations of entire countries (e.g., National Council of Taechers of Mathematics, 2000).

As in the other areas addressed in this book, Alan Bishop's early research on spatial abilities and visualization in mathematics education has provided insights and a foundation for research that is ongoing after almost three decades. Thus the continuity implied in synechism is alive and well in this area, and Bishop will be remembered for his contribution.

References

Anderson, M., Sáenz-Ludlow, A., Zellweger, Z., & Cifarelli, V. V. (Eds.). (2003). *Educational perspectives on mathematics as semiosis: From thinking to interpreting to knowing.* New York: Legas.

Aspinwall, L., Shaw. K. L., & Presmeg, N. C. (1997). Uncontrollable mental imagery: Graphical connections between a function and its derivative. *Educational Studies in Mathematics, 33*(3), 301–317.

Bishop, A. J. (1980). Spatial abilities and mathematics education – a review. *Educational Studies in Mathematics, 11,* 257–269.

Bishop, A. J. (1983). Space and geometry. In R. Lesh & M. Landau (Eds.), *Acquistion of mathematical concepts and processes* (pp. 175–203). New York: Academic Press.

Bishop, A. J. (1988a). *Mathematical enculturation: A cultural perspective on mathematics education*. Dordrecht, The Netherlands: Kluwer Academic Publishers.

Bishop, A. J. (1988b). Mathematics education in its cultural context. *Educational Studies in Mathematics, 19,* 179–191. [Reprinted 2004 as Chapter 16 in T. P. Carpenter, J. A. Dossey, and J. L. Koehler (Eds.), *Classics in mathematics education research* (pp. 201–207). Reston, Virginia: National Council of Teachers on Mathematics.]

Bishop, A. J., (1989). Review of research on visualization in mathematics education. *Focus on Learning Problems in Mathematics, 11*(1), 7–16.

Bishop, A. J., Clements, K., Keitel, C., Kilpatrick, J., & Laborde, C. (Eds.). (1996). *International handbook of mathematics education.* Dordrecht: Kluwer Academic Publishers.

Clements, M. A. (1978). Some mathematics educators' view on spatial ability. In B. Low, M. E. Dunkley, & J. Conroy (Eds.), *Research in mathematics education in Australia, 1978.* Sydney, Australia: Macquarrie University.

Cobb, P. (2007). Putting philosophy to work. In F. K. Lester, Jr. (Ed.), *Second handbook of research in mathematics teaching and learning* (pp. 3–38). Charlotte, NC: Information Age Publishing.

Denzin, N. K., & Lincoln, Y. S. (Eds.). (2000). *Handbook of qualitative research.* Thousand Oaks, California: Sage Publications.

English, L. D. (Ed.) (2002). *Handbook of international research in mathematics education.* Mahwah, New Jersey: Lawrence Erlbaum Associates.

Goldin, G. A., & Janvier, C. (Eds.). (1998). Representations and the psychology of mathematics education. Special double issue of *The Journal of Mathematical Behavior, 11 & 12.*

Gravemeijer, K., Lehrer, R., van Oers, B., & Verschaffel, L. (Eds.). (2002). *Symbolizing, modeling and tool use in mathematics education.* Dordrecht, The Netherlands: Kluwer Academic Publishers.

Gray, E. (1999). Spatial strategies and visualization. In O. Zaslavsky (Ed.), *Proceedings of the 23rd Conference of the International Group for the Psychology of Mathematics Education, 1,* 235–242.

Grouws, D. A. (Ed.) (1992). *Handbook of research on mathematics teaching and learning.* New York: Macmillan.

Gutierrez, A., Jaime, A., & Fortuny, J. M. (1991). As alternative paradigm to evaluate the acquisition of the van Hiele levels. *Journal for Research in Mathematics Education, 22*(3), 231–251.

Hitt, F. (Ed.). (2002). *Representations and mathematics visualization.* Mexico City: Cinvestav-IPN.

Hostetler, K. (2005). What is "good" educational research? *Educational Researcher, 34*(6), 16–21.

Johnson, R. K., & Onwuegbuzie, A. J. (2004). Mixed methods research: A research paradigm whose time has come. *Educational Researcher, 33*(7), 14–26.

Kelly, A. E., & Lesh, R. A. (Eds.). (2000). *Handbook or research design in mathematics and science education.* Mahwah, New Jersey: Lawrence Erlbaum Associates.

Kirshner, D., & Whitson, J. A. (Eds.). (1997). *Situated cognition: Social, semiotic, and psychological perspectives.* Mahwah, New Jersey: Lawrence Erlbaum Associates.

Krutetskii, V. A. (1969). Papers in J. Kilpatrick & I. Wirszup (Eds.), *Soviet studies in mathematics education: Vol. II, The structure of mathematical abilities.* Chicago: University of Chicago Press.

Krutetskii, V. A. (1976). *The psychology of mathematical abilities in school children.* Edited by J. Kilpatrick & I. Wirszup. Chicago: University of Chicago Press.

Lave, J. (1997). The culture of acquisition and the practice of understanding. In D. Kirshner & J. A. Whitson (Eds.), *Situated cognition: Social, semiotic, and psychological perspectives* (pp. 17–35). Mahwah, New Jersey: Lawrence Erlbaum Associates.

Lean, G. A. (1981). *Spatial training studies and mathematics education: A review.* Unpublished paper, University of Cambridge.

Lean, G. A., & Clements, M. A. (1981). Spatial ability, visual imagery and mathematical performance. *Educational Studies in Mathematics, 12*, 267–299.

Lehrer, R., & Pritchard, C. (2002). Symbolizing space into being. In K. Gravemeijer, R. Lehrer, B. van Oers, & L. Verschaffel (Eds.), *Symbolizing, modeling and tool use in mathematics education* (pp. 59–86). Dordrecht, The Netherlands: Kluwer Academic Publishers.

Lester, F. K., Jr. (Ed.). (2007). *Second handbook of research in mathematics teaching and learning*. Charlotte, NC: Information Age Publishing.

Mariotti, M. A. (2002). Influences of technologies' advances on students' mathematics learning. In L. D. English (Ed.), *Handbook of international research in mathematics education* (pp. 695–723). Mahwah, New Jersey: Lawrence Erlbaum Associates.

National Council of Teachers of Mathematics (2000). *Principles and standards for school mathematics*. Reston, Virginia: The Council.

Neisser, U. (1970). Visual image as process and as experience. In J. S. Antrobus (Ed.), *Cognition and effect*. Boston: Little & Brown.

Owens, K. (1999). The role of visualisation in young students' learning. In O. Zaslavsky (Ed.), *Proceedings of the 23rd Conference of the International Group for the Psychology of Mathematics Education, 1*, pp. 220–234.

Parzysz, B. (1999). Visualization and modeling in problem solving: From algebra to geometry and back. In O. Zaslavsky (Ed.), *Proceedings of the 23rd Conference of the International Group for the Psychology of Mathematics Education, 1*, pp. 212–219.

Peirce, C. S. (1992). *The essential Peirce* (vol. 1) Edited by N. Houser & C. Kloesel. Bloomington: Indiana University Press.

Peirce, C. S. (1998). *The essential Peirce* (vol. 2) Edited by the Peirce Edition Project. Bloomington: Indiana University Press.

Presmeg, N. C. (1985). *The role of visually mediated processes in high school mathematics: A classroom investigation*. Unpublished Ph.D. dissertation, University of Cambridge.

Presmeg, N. C. (1992). Prototypes, metaphors, metonymies, and imaginative rationality in high school mathematics. *Educational Studies in Mathematics, 23*, 595–610.

Presmeg, N. C. (1997). Generalization using imagery in mathematics. In L. D. English (Ed.), *Mathematical reasoning: Analogies, metaphors and images* (pp. 299–312). Mahwah, New Jersey: Lawrence Erlbaum Associates.

Presmeg, N. C. (2006a). Research on visualization in learning and teaching mathematics. In A. Gutiérrez & P. Boero (Eds.), *Handbook of research on the psychology of mathematics education: Past, present and future* (pp. 205–235). Rotterdam, The Netherlands: Sense Publishers.

Presmeg, N. C. (2006b). A semiotic view of the role of imagery and inscriptions in mathematics teaching and learning. Plenary Paper. In J. Novotna, H. Moraova, M. Kratka, & N. Stehlikova (Eds.), *Proceedings of the 30th Annual Meeting of the International Group for the Psychology of Mathematics Education, 1*, 19–34. Prague, July 16–21, 2006.

Richardson, A. (1969). *Mental imagery*. London: Routledge & Kegan Paul.

Richardson, A. (1972). Voluntary control of the memory image. In P. W. Sheehan (Ed.), *The function and nature of imagery*. New York: Academic Press.

Shepard, R. N., & Metzler, J. (1971). Mental rotation of three dimensional objects. *Science, 171*, 701–703.

Steffe, L. P., & Thompson, P. W. (2000). Teaching experiment methodology: Underlying principles and essential elements. In Kelly, A. E. & Lesh, R. A. (Eds.), *Handbook or research design in mathematics and science education* (pp. 267–306). Mahwah, New Jersey: Lawrence Erlbaum Associates.

Wheatley, G. H. (1997). Reasoning with images in mathematical activity. In L. D. English (Ed.), *Mathematical reasoning: Analogies, metaphors and images* (pp. 281–297). Mahwah, New Jersey: Lawrence Erlbaum Associates.

Yerushalmy, M., Shternberg, B., & Gilead, S. (1999). Visualization as a vehicle for meaningful problem solving in algebra. In O. Zaslavsky (Ed.), *Proceedings of the 23rd conference of the international group for the psychology of mathematics education, 1*, pp. 197–211.

Chapter 7
Spatial Abilities, Mathematics, Culture, and the Papua New Guinea Experience

M.A. (Ken) Clements

Sometime in 1976 Dr. Peter Fensham, then Professor of Science Education at Monash University in Melbourne, Australia, asked me if I would like to nominate a distinguished mathematics educator who might be invited to work as a Visiting Fellow for a period of up to 6 months within Monash University's Faculty of Education. At that time I led Monash University's large mathematics teacher education program. There had been a number of distinguished science education visitors to Monash Education in the mid-1970s, and Dr. Fensham felt that it was "mathematics education's turn."

In 1976 I was a young mathematics educator, in only my third year as an academician at Monash University. I had never visited Europe, or America, and had never corresponded with, or had any form of close contact with any distinguished overseas mathematics educator. However, I was in the habit of avidly reading the main mathematics education research journals and periodicals, particularly those published by the National Council of Teachers of Mathematics in America (Journal for Research in Mathematics Education, The Mathematics Teacher, and The Arithmetic Teacher), and the two main English mathematics education journals Mathematical Gazette and Mathematics Teaching. I also read Educational Studies in Mathematics, a journal which, at that time, was published in the Netherlands by D. Reidel and edited by Dr. Hans Freudenthal, a Dutch mathematician/educator.

It says something about Alan Bishop's work up to that time that, before I had ever met him, the first name that sprang to my mind for the position of Visiting Fellow was his. I knew he worked at Cambridge University, and I had read quite a lot of articles that he had written for Mathematics Teaching. In particular, I had been impressed by his (1972) article, "Trends in Research in Mathematics Education", which had provided a succinct but highly informative summary of what was going on in mathematics education research in the United States, in England, and in Continental Europe. For example, in the third paragraph of Bishop's (1972) article he had contrasted the writings of Jerome Bruner, Jean Piaget, Lee Shulman and

M.A. (Ken) Clements
Department of Mathematics, Illinois State University, Normal, IL 61790-4520, United States of America
e-mail: clements@ilstu.edu

P. Clarkson, N. Presmeg (eds.), *Critical Issues in Mathematics Education*,

Zoltan Dienes. His main focus seemed to be on research into the development of curricula that featured links between school mathematics and the wider community (Bishop & McIntyre, 1969, 1970).

I had noted that in his writings Bishop often raised the question of which criteria should be used to evaluate mathematics curricula. He obviously had an interest in educational philosophy, and was influenced by the work of his Cambridge colleague, Professor Paul Hirst. In his 1972 article he suggested that much of Piaget's work was "uncontrolled", and gave a positive nod in the direction of Skemp's "schematic learning" approaches (p. 15). He had commented that Bloom's (1956) Taxonomy impressed testers more than teachers. In the course of describing the work of David Wheeler, Bishop (1972) had looked forward to an era in which mathematics would be seen as an invention rather than as a discovery, and to a time when it would be recognized that "the teacher is not the builder – the child is" (p. 16).

In the same 1972 article Bishop surveyed research within the United Kingdom and on the Continent into "the teacher and his method" (p. 16). He drew special attention to research on teachers' use of structured and visual materials, and referred to investigations that he had carried out with Frank Land (Land & Bishop, 1969). He mentioned that some of his most recent research had been highly "value-laden in its goals" (p. 16), insofar as it had largely been concerned with matters related to a child's independence, autonomy, self-awareness, self-reliant thinking. He maintained that, whereas U.S. mathematics education researchers were tending to quantify classroom-interaction data, in the United Kingdom the focus tended to be on the "quality of the interaction" (p. 16). He also referred to his research on teachers' decision making in the classroom (Bishop, 1970). At the end of his research overview he pointed out that although much European mathematics education research was less concerned with developing theory and was more concerned with aspects of curriculum development and classroom interaction, "there is nothing as practical as a good theory" (p. 17).

In view of the fact that, from 1977 onwards, Alan Bishop would make rich contributions to thinking about the influence on mathematics teaching and learning of macro- and micro-cultural settings, and to how personal values held by individuals influenced mathematics curriculum and teaching decisions, it is intriguing to note that despite the fact that his 1972 overview was prepared for a UNESCO publication on "new trends in teaching mathematics", the word "culture" did not appear in the whole article. Nor did the article contain any variant (e.g., "cultural", or "enculturation") of that word. The word "values" appeared on just two occasions, both in relation to values implicit in classroom decisions made by teachers.

Because I had been a secondary mathematics teacher for 10 years before taking up my position at Monash University, Bishop's obvious emphasis on classroom-relevant research made his writings particularly interesting for me. I also shared his interest in possible relationships between spatial – or visual – abilities and mathematics thinking. In his 1972 paper he had referred, positively, to the work of MacFarlane Smith (1964), who had developed a series of highly regarded spatial tests and had carried out a series of experiments concerned with visual abilities and mathematics teaching and learning. MacFarlane Smith had claimed that spatial ability and mathematical ability were highly correlated and that his research indicated that these abilities could be developed by carefully chosen structured materials. According to Bishop (1972), this was "one of the main benefits accruing from the use of structured materials in primary schools" (p. 16). I had also read Bishop's (1973) speculative paper on possible relationships between teachers' and students' use of structural apparatus in mathematics classrooms and students' development of spatial abilities. Indeed, that paper had motivated me to conduct, and supervise, research on links between spatial abilities and mathematics learning (see, e.g., Wattanawaha & Clements, 1982). In that context, I was aware of Krutetskii's (1976) claim that a preference for using visual, as opposed to verbal-logical, thinking, was not necessarily an essential characteristic of a mathematically talented person. I also knew that Bishop was well aware of Krutetskii's conclusions on that matter (Bishop, 1976).

It was not surprising, therefore, that the idea of having Alan Bishop at Monash University, even if it would be for only 6 months, was extremely attractive to me. You can imagine my delight when, sometime toward the end of 1976, Peter Fensham told me that Alan Bishop had accepted Monash University's invitation to come to Monash University between July and December, 1977. Peter told me that I should prepare well for the visit, in order that I would derive as much benefit as possible from it.

As Alan's current position was at Cambridge University, I expected he would be a rather crusty don, with a deep scholarly interest in all things cerebral. The latter proved to be true, but I did not expect to find him a fit and very active young man (about 40, but looking much younger), with exceptional gifts in music. He was accompanied to Melbourne by his linguistically talented wife (Jenny) and their two precocious and lively boys (Simon and Jason). Alan had sent careful instructions ahead that he wanted the boys to attend a government primary school as close to Monash University as possible. My preconceived ideas of his being someone who would demand an aristocratic lifestyle for himself and his family were clearly wide of the mark.

Alan fitted into life at Monash, quickly and seamlessly. He taught a masters-level class in "Spatial Abilities and the Curriculum" which I – and Nongnuch Wattanawaha, my master's student who was researching spatial ability and mathematics learning – attended. We quickly found that his views on spatial abilities and school mathematics differed significantly from our own. He thought of "spatial abilities", in the plural, whereas we thought of spatial ability and visualization as those abilities which had been identified and defined by factor analysts working largely within psychological or psychometric traditions. He was decidedly interested in the imagery that might be developed as a consequence of the use of a variety of structured learning aids, whereas we had not yet linked spatial ability so carefully with curriculum materials and teaching approaches. He was especially interested in the thinking processes people used when tackling mathematical tasks. Those emphases were new to me, but as a result of Alan's contagious enthusiasm I would subsequently become interested in and influenced by them.

I had not anticipated that Alan, and his family, would visit Papua New Guinea (PNG), for 3 months, en route to Monash University. I quickly learned, from Alan, that his time in PNG had challenged most of his previous assumptions and thinking about education and schooling in general, and about mathematics and mathematics education, in particular. On his visit to PNG he had befriended Glendon ("Glen") Lean, an Australian scholar who was researching the indigenous counting systems of Oceania, Micronesia and Polynesia. Undoubtedly, Alan's efforts to comprehend what was going on in the name of school mathematics in Papua New Guinea were profoundly influenced and enhanced by his association with Glen, who, at that time, was Acting Director of the new Mathematics Education Centre at the PNG University of Technology (UNITECH), in the city of Lae. Alan's work with Glen would bear much fruit. Alan became Glen's doctoral supervisor, and in the 1990s Glen completed his pioneering and seminal doctoral thesis on the indigenous counting systems of PNG and Oceania (Lean, 1992).

In what remains in this chapter I will briefly tell a story of some of the aspects of Alan's work during the period 1977–1980 which, I believe, were crucially important in his subsequent growth as a scholar. The chapter will focus on how his developing ideas on the role of spatial abilities and visualization in school mathematics would have a large impact on how he would come to theorise the domain of "mathematical enculturation." I will argue that his 1977 visit to PNG would change the way he thought about, conducted, supervised, wrote about, and sought opportunities to carry out mathematics education research.

Is a Picture Worth a Thousand Words?

The title of the article which Alan wrote, for his regular "Research" column, for Volume 81 of Mathematics Teaching (Bishop, 1977) was "Is a Picture Worth a Thousand Words?" The article was written while Alan was at Monash University, and was largely influenced by his analyses of the interview data he had collected from 12 first-year students at UNITECH in PNG, just before he came to Monash University. I believe that this article announced to the world a new focus for his research into mathematics teaching and learning. From now on, he seemed to be saying, the influence of culture would be an important consideration in any research that he would plan, conduct, and report.

Bishop began his 1977 article by saying that "one of the most difficult things for us to realize is the extent of our own knowledge and learning" and that "the peculiar nature of knowledge is that once you have it you cannot regain the state of not having it" (p. 32). He went on to show how so much of what is communicated by pictures depends on the cultural backgrounds of those who are attempting to "read" the pictures. He then commented that the difficulties "people from non-Western cultures have with certain Western ideas and representations are becoming well documented and the literature contains many speculations regarding the causes" (p. 33). After taking a jab at cross-cultural researchers who relied almost solely on analyses of responses to items on pencil-and-paper survey instruments, he went on to outline the research procedures he had used at UNITECH.

His UNITECH interviews had occupied more than 80 hours altogether, and had been carried out on a one-to-one basis. The tasks that he had employed in the interviews were extraordinarily creative. He had found some of them in the cross-cultural education literature (e.g., Deregowski & Munro, 1974; Kearins, 1976), and had created others himself. He had developed a 3 by 3 grid model for classifying tasks and responses to tasks: each mode of presentation for a task could be classified into one of three categories (word symbols, diagrams or photographs, objects), and each mode of response could be regarded as belonging to one of the same three categories. Thus, the task of making objects shown in a diagram would be regarded as a "diagrams-objects" task. In the paper, Alan argued that Western conventions were often used in pictures to indicate sequence, or depth, and that these conventions needed to be learned. Thus, for example, children in non-Western cultures did not naturally grow to realize that dotted lines in a line drawing of a cube can indicate edges "at the back" which cannot be seen from a front view. That is a Western convention, and is something which is learned, often as a result of specific instruction. He commented that much of Western mathematics involved conventions that were not naturally acquired but could be learned as a result of training. He added that analyses of his PNG data indicated that often students who had grown up and had attended community schools in PNG villages were not aware of, and therefore had not learned, many standard mathematics conventions.

Towards the end of his paper Bishop (1977) stated that he had uncovered some strengths among his UNITECH interviewees. He described one such strength in relation to the following task:

> Set out 12 small objects (pencil, coin, paper clip, etc.) in a 4 × 3 array; let your friend look at it for 45 seconds, mess up the arrangement, and ask your friend to replace the pieces in the same position as they were originally.
>
> (p. 35)

The 12 students whom Alan interviewed rarely made a mistake on that task. Then the same kind of thing was repeated, only this time 12 playing cards were used instead of 12 everyday objects. In a third form of the task, 12 local objects (feathers, shells, etc.) were used. Alan said his 12 interviewees were "excellent" on these tasks.

Shortly after Bishop's time at Monash University, complete reports on his PNG research became available (Bishop, 1978a, 1978b). As a result of reflections on his PNG data, Bishop (1978a) put forward the following 12 researchable hypotheses:

1. Pictorial representation conventions are teachable.
2. Training students in drawing and sketching techniques improves their ability to read and interpret other people's diagrams.
3. Manipulative work with concrete apparatus aids and improves students' spatial abilities.
4. There are significant variations in the capacity of local languages to express social ideas.
5. The variation in quality of local spatial vocabulary accompanies visualizing skill.

6. Orientation and mapping skills are more developed in PNG students than other types of spatial abilities.
7. Memory tasks are performed by PNG students with little or no verbal mediation.
8. Within-group variability in spatial ability is greater than between-group variability.
9. Less acculturated students have better visual memory than students who are more acculturated.
10. Students coming from areas where the local languages contain no (easy) conditional mood will tend towards a greater use of visual memory and ikonic processing.
11. Teaching strategies emphasizing "understanding" will be less successful in the short term than those that emphasize "memory".
12. Symbolic and hierarchical processing and coding is teachable. (pp. 36–40).

Implicit in these conjectures was a belief that PNG students tended to have good memories but, relative to Western students, found it difficult to grasp abstract principles. In commenting on his eleventh researchable hypothesis (above), Bishop (1978a) wrote:

> From my perspective the whole educational development exercise is enmeshed in the long slow process of cultural adaptation. Strategies which are designed to foster understanding, the "meaning" of general principles, the subtle use of example and counter-example to extend or test out generalizations and hypotheses, etc., are so dependent on the supporting framework of "Western" ideals, philosophies and societal values that they are literally meaningless within the PNG culture as it is at present.
>
> (p. 39)

Bishop added that teaching strategies should be devised which relied on and fostered strengths. He argued that, whether one liked it or not, it was necessary for PNG curriculum planners and teachers to recognize that PNG students were not ready for highly theoretical analyses of content, and, that this was especially true of community (i.e., primary) school children. Levy-Bruhl (1966, quoted in Cole & Scribner, 1974) argued in a similar way. Lancy (1978), on the other hand, argued more along the lines of Cole and Scribner (1974), that "we are unlikely to find cultural differences in basic component cognitive processes", and that "there is no evidence in any line of investigation that we have reviewed that any cultural group lacks a basic process such as abstraction, or inferential reasoning or categorization" (p. 193). Data supporting that point of view, but based on studies conducted outside of Papua New Guinea, were presented around the same time that Bishop was writing, by Stevenson, Parker, Wilkinson, Bonnevaux, and Gonzales (1978), Sharp, Cole, and Lave (1979), and Kagan, Klein, Finley, Rogoff, and Nolan (1979).

While at Monash, Alan told me of the internal cognitive conflicts he was experiencing as he struggled to come to grips with the challenges of his PNG data. He had never before realized that a student's cultural and social surroundings, language, and spatial preferences, could interact so profoundly with ways in which mathematics was presented, taught, and understood. It seemed to me, at the time, that he believed that he had been passed a baton of responsibility to communicate to others the lessons he was learning.

The Call to be Editor of *Educational Studies in Mathematics*

One day, during Alan Bishop's 6 months at Monash University in 1977, he showed me a letter he had just received from Dr. Hans Freudenthal. On several occasions, Alan had spoken to me of his admiration for the way Freudenthal dared to be different in mathematics education. He saw Freudenthal as someone who had not been afraid to challenge many of the received traditions about the teaching and learning of mathematics. Alan had often told me that, in particular, Freudenthal, the foundation editor of *Educational Studies in Mathematics*, did not like many of the trends in American mathematics education research (see, e.g., Freudenthal, 1979). Furthermore, Bishop recognized that Freudenthal had been prepared to risk the ire of European educators by courageously attacking some of the ideas that Jean Piaget had put forward regarding mathematics and mathematics education (see, e.g., Freudenthal, 1973).

The letter from Freudenthal asked Alan if he would be willing to become editor of *Educational Studies in Mathematics* (ESM). Alan recognized immediately the huge honour that Freudenthal was bestowing on him, and of course was keen to accept the invitation. However, at Cambridge University he was usually under considerable pressure to meet the many demands placed on him there. Furthermore, he quietly confided to me, one day he hoped to be appointed to a Chair in Education. Would accepting Freudenthal's invitation to become editor of ESM result in his having bitten off more than he could chew? My answer was quick and to the point. I told him that he was perfectly placed to succeed Freudenthal as editor. His Cambridge appointment would certainly be consistent with his position as editor and, so far as possible future promotion to a Chair was concerned, holding the position of editor of ESM would certainly do him no harm. Subsequently, history would reveal that Alan would become the second editor of ESM, that his work in that role would lift the journal to new heights, and that one day he would return "downunder" to take up the position of Professor of Education at Monash University.

A Personal Dilemma

Having accepted the position of editor of ESM, Alan was faced with a personal dilemma. Analysing and interpreting his PNG data had been, perhaps, the most significant and challenging intellectual exercise of his academic life to that time. He really wanted the world to reflect on some of the issues with which he had been wrestling. Would it be professionally appropriate for him to use ESM as a vehicle for one or two articles in which he set out the ideas he had expressed in his PNG articles (Bishop, 1977, 1978a, 1978b)? I understood his concern, especially because he was just taking over the position of editor. I advised him to write his articles and submit them to independent reviewers – that is to say, his papers should be subjected to exactly the same peer review scrutiny as other papers submitted to ESM. Alan took my advice, and as a result two major, transformative papers appeared in ESM (Bishop, 1979, 1980).

Concluding Comments

It would be wrong to complete this paper without mentioning that Alan Bishop has had an enormous influence on my own thinking about mathematics education. During the second half of 1977 he and I ran workshops for mathematics teachers in far-flung parts of the State of Victoria, Australia, and on those occasions I was struck by his determination to involve participants actively in the learning process. In his head he carried an extremely rich set of mathematics learning activities. But, after his PNG visit he was reluctant to use the activities too much. He explained that his focus had moved towards meeting individual students and teachers where they were, "now". He was less interested in showing others what brilliant activities he could introduce into workshop situations.

In his later, classic book Mathematical Enculturation, Alan Bishop (1988) would subsequently write:

> It should be clear by now that, fundamentally, education for me must be recognized as being a social process, and therefore a mathematics education must also have at its core the assumption of being a social process. It seems so trivial to say this, and yet . . . the human, the essentially interpersonal nature of education is so often ignored in the rush for the acquisition of mathematical techniques and the desire for so-called efficiency in mathematics education.
>
> If we therefore consider these social aspects of mathematics education, we find that there are five significant levels of scale involved:
>
> Cultural
> Societal
> Institutional
> Pedagogical
> Individual
>
> The largest social group is the cultural group and mathematics as a cultural phenomenon is clearly supra-societal in nature. Mathematics is used in every society; mathematics is the only subject taught in most schools in the world. . . .
>
> (pp. 13–14)

This was a message that Alan Bishop, perhaps more than anyone else, would bring to the world of mathematics education for the next quarter of a century – and beyond that. It was, and still is, the most important message of all if we are to make school mathematics more accessible, more equitable, more stimulating, and ultimately more worthwhile for all students. No-one has had more influence in pointing us in the right directions, and in leading us to some of the many hidden pathways, than Alan Bishop.

A postscript: I should add that subsequently I too had the privilege of working with Glen Lean in PNG, for the first 7 months of 1980 and again several years later.

Between September and November 1980 I was able to spend time at the Institute of Education at Cambridge University, and discovered that Alan had established a graduate mathematics education program in which students from all over the world were participating. Not surprisingly, cultural factors had become the new imperative in his courses. At the same time as I was at Cambridge, Glendon Lean was also there, and so too was Professor Peter Jones, who had been at UNITECH throughout Alan's time in PNG in 1977. Alan took the opportunity to organize a special "PNG Mathematics Seminar," to which we all contributed sessions, including Alan himself. About 40 people came from far and wide to attend the Seminar, and it was a thrill for me to realise that what Alan had learned in PNG was so highly valued, even within the hallowed halls of learning at Cambridge University.

References

Bishop, A. J. (1970). Stimulating pedagogical decision making. *Visual Education, 1970*, 41–43.

Bishop, A. J. (1972). Trends in research in mathematics education. *Mathematics Teaching, 58*, 14–17.

Bishop, A. J. (1973). Use of structural apparatus and spatial ability: A possible relationship. *Research in Education, 9*, 43–49.

Bishop, A. J. (1976). Krutetskii on mathematical ability. *Mathematics Teaching, 77*, 31–34.

Bishop, A. J. (1977). Is a picture worth a thousand words? *Mathematics Teaching, 81*, 32–35.

Bishop, A. J. (1978a). *Spatial abilities in Papua New Guinea*. Mathematics Education Centre Report No. 2. Lae, Papua New Guinea: Papua New Guinea University of Technology.

Bishop, A. J. (1978b). *Visualizing and mathematics in a pre-technological culture*. Mathematics Education Centre Report No. 4. Lae, Papua New Guinea: Papua New Guinea University of Technology.

Bishop, A. J. (1979). Visualizing and mathematics in a pre-technological culture. *Educational Studies in Mathematics, 10*(2), 135–146.

Bishop, A. J. (1980). Spatial abilities and mathematics education – A review. *Educational Studies in Mathematics, 11*, 257–269.

Bishop, A. (1988). *Mathematical enculturation: A cultural perspective on mathematics education*. Dordrecht, the Netherlands: Kluwer Academic Publishers.

Bishop, A. J. & McIntyre, D. I. (1969). A comparison of secondary and primary teachers' opinions regarding the content of primary school mathematics. *Primary Mathematics, 7*(2), 33–41.

Bishop, A. J. & McIntyre, D. I. (1970). A comparison of teachers' and employers' opinions regarding the content of secondary school mathematics. *Mathematical Gazette, 54*, 229–233.

Bloom, B. S. (1956). *Taxonomy of educational objectives – Cognitive domain*. London: Longmans.

Cole, M. & Scribner, S. (1974). *Culture and thought*. New York: John Wiley & Sons.

Deregowski, J. B. & Munro, D. (1974). An analysis of polyphasic pictorial perception. *Journal of Cross-Cultural Psychology, 5*(3), 329–343.

Freudenthal, H. (1973). *Mathematics as an educational task*. Dordrecht, the Netherlands: D. Reidel Publishing Co.

Freudenthal, H. (1979). Ways to report on empirical research in education. *Educational Studies in Mathematics, 10*, 275–303.

Kagan, J., Klein, R. E., Finley, G. E., Rogoff, B., & Nolan, E. (1979). A cross-cultural study of cognitive development. *Monographs of the Society for Research in Child Development, 44*(5).

Kearins, J. (1976). Skills of desert children. In G. E. Kearney & J. McElwain (Eds.), *Aboriginal cognition: Retrospect and prospect* (pp. 54–66). New Jersey: Humanities Press.

Krutetskii, V. A. (1976). *The psychology of mathematical abilities in school-children* (translated by J. Teller and edited by J. Kilpatrick). Chicago: University of Chicago Press.

Lancy, D. F. (1978). Cognitive testing in the Indigenous Mathematics Project. *Papua New Guinea Journal of Education, 14*, 14–142.

Land, F. W. & Bishop, A. J. (1969). *Reports of mathematics teaching research project*. University of Hull Institute of Education.

Lean, G. A. (1992). *Counting systems of Papua New Guinea and Oceania*. Unpublished PhD thesis, Papua New Guinea University of Technology (Lae, PNG).

MacFarlane Smith, I. (1964). *Spatial ability: Its educational and social significance*. London: University of London Press.

Sharp, D., Cole, M., & Lave, C. (1979). Education and cognitive development: The evidence from experimental research. *Monographs of the Society for Research in Child Development, 44*(1–2).

Stevenson, H. W., Parker, T., Wilkinson, A., Bonnevaux, B., & Gonzales, M. (1978). Schooling, environment and cognitive development: A cross-cultural study. *Monographs of the Society for Research in Child Development, 43*(2).

Wattanawaha, N. & Clements, M. A. (1982). Qualitative aspects of sex-related differences in performance on pencil-and-paper spatial questions, grades 7–9. *Journal of Educational Psychology, 74*, 878–887.

Section IV
Cultural and Social Aspects

It is not a mistake that the key article for this section has in its title "visualising". It could be thought that this article should be placed in the previous section. And it could be, for all sections in this volume overlap to some degree. But this article signals a dramatic shift for Bishop's research. In this article he recounts some of the work carried out in Papua New Guinea, which he found so confronting. He changed.

Another reading of this article points in a slightly different direction. Bishop seems to be forced back to consider again in more depth issues he had contained in the rectangles when he was conceptualising in the 1970s how teachers came to the quick but insightful decisions in their classrooms in the middle of the act of teaching (see Fig. 1 [p. 28] in the introduction to Section II, which also appears as Fig. 4.1 [p. 40]). But from the time of his Papua New Guinea visit, exploring the implications of social and cultural (and political) issues, acting both inside and outside the classroom, would be a decisive refocusing for Bishop's research agenda. An important outcome of this thinking is Bishop's oft quoted book published in 1988.

Barton, reflecting on Bishop's contribution to the opening up of mathematics education research to the sociocultural field, reminds us that Bishop also contributed important early ideas in the conceptualisation of what we know today as ethnomathematics. Leung starts with Bishop's article and uses it to begin a discussion of mathematics education in two cultural situations that are quite different to that of the west, and indeed that of Papua New Guinea the cultural contexts that had so confronted Bishop. Leung considers East Asian societies that draw upon the Confucian cultural tradition, and then the Near Eastern societies that draw on an Islamic tradition. Leung leaves it to the reader to contemplate how their culture enculturates their students into the ways of doing and understanding mathematics.

Additional Bishop References Pertinent to This Issue

Bishop, A. J. (1985). The social construction of meaning – a significant development for mathematics education. *For the Learning of Mathematics, 5*(1), 24–28.

Bishop, A. J. (1988). *Mathematical enculturation: A cultural perspective on mathematics education*. Dordrecht: Kluwer.

Chapter 8
Visualising and Mathematics in a Pre-Technological Culture*

Alan J. Bishop

Papua New Guinea is a strange, fascinating country which is at present going through an amazing period of change. All countries experience change, but it is possible that few have ever experienced change so rapid as that in Papua New Guinea (PNG). "From stone-age to twentieth century in one lifetime" is no overstatement. Apart from those living in the few small towns the majority of the population have little contact with the technological society and culture which we know so well. And yet, there are two universities, and I was fortunate to be able to spend three months last year working at one of them, the University of Technology at Lae. (The other is at Port Moresby, the capital.)

In any culture it is likely that one will find a few people who possess certain skills naturally, one might say, whether or not that culture *prizes* those skills. For example, we treat as amazing oddities those individuals who have exceptional memories – they are often given entertainer status, and are *not* awarded the same respect as they would be in Papua New Guinea. The 'big-men' there have, among other attributes, exceptional memories (Strathern, 1977).

The inverse is that in Papua New Guinea there will be some individuals who possess those skills necessary for doing mathematical and scientific work. Some of these have been found, and at the University of Technology there are a few 'local' tutors employed to teach. But in a pre-technological culture these skills are rare, their worth is not appreciated and their presence is not even recognised. In these conditions teaching mathematics and science is far more complex than it is in most technological cultures.

But this article is not concerned with the strategies of educational development in Papua New Guinea. Rather I want to describe some of the data from my research there and to encourage you to consider what might be the implications for the learning and teaching of mathematics in your own cultures and countries.

There is a second concern. It is all too easy when reading descriptions of research, to generalise, and this is particularly the case in the field of mathematics education. It is almost as difficult for me to stop generalising as it is for some people in certain

*A version of this article was presented as a paper to the Second International Conference for the Psychology of Mathematics Education held in September 1978 at Osnabruck, W. Germany.

Educational Studies in Mathematics 10(1979) 135–146. 0013–1954/79/0102–0135$01.20

cultures to start generalising. In a journal such as this, with an international readership, it is very important to recognise that as well as sharing common research interests there are many differences between us. In particular, what may be the case in one country or culture may be quite different elsewhere. My data from Papua New Guinea will, I hope, be a strong reminder of this.

My research there was concerned with the visual and spatial aspects of mathematics and was an extension of work which I have been carrying out for several years (Bishop, 1974). I was testing in great detail, twelve male first year University students whose ages varied from 16 to 26. The aim was to identify relative strengths and weaknesses in the spatial field, and to attempt to relate these to the different linguistic, environmental and cultural features of the students' background. Accordingly, the students were carefully chosen on a variety of criteria and they were drawn from three specific areas of the country: the Capital, a Highland region (Enga) and an Island region (Manus). They were studying a variety of courses, but all were entering a field of technology, e.g., engineering, agriculture, architecture, cartography, accountancy, etc.

The testing used approximately forty different tasks and was carried out individually in my University office. The language used was English, and although for some it was their third or even fourth language they were all fluent in it. (English is the 'academic' language used there although Pidgin and Motu have a generally wider currency.) As I was able to obtain 6 to 7 hours individual test data from each of the twelve students, it would not be appropriate to describe all the details here. Two major reports have been written on the work (Bishop, 1977, 1978), but in this article I will summarize the main ideas and give a few specific examples from my data, and occasionally from other data, to illustrate the main points. My comments will be grouped under five main headings: Picture Conventions, Drawing, Visualising, Language, and Cognitive Characteristics.

Picture Conventions

It was made clear by several tasks that there existed a general unfamiliarity with many of the conventions and 'vocabulary' of the diagrams commonly used in Western education and which are now entering PNG schools. Some tasks showed this dramatically because they focussed directly on the convention. For example, the students were asked to make models using plasticine "corners" and cocktail sticks, based on drawings. The drawings used were similar to those used by Deregowski (1974) and cues such as shaping and dotted lines were used to indicate depth.

The representation of a three-dimensional object by means of a two-dimensional diagram demands considerable conventionalising which is by no means immediately recognisable by those from non-Western cultures. Two of the Highland students produced perfectly flat, 2D objects when shown the diagrams in Figure 8.1.

It is clear that these students were unfamiliar with the oblique convention where the front (square) of the object (cube) is drawn and the rest displaced from it. Only

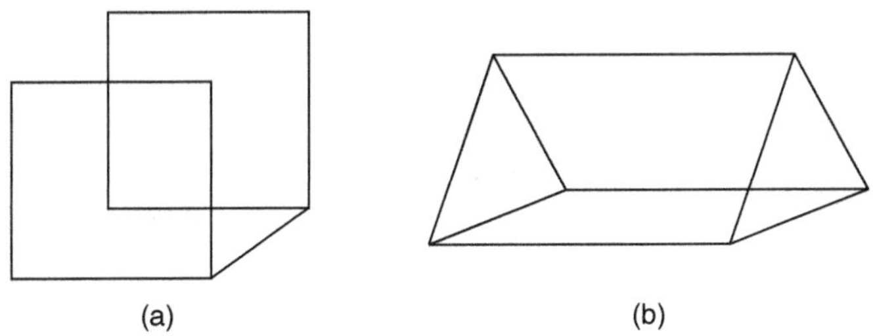

Fig. 8.1.

the Manus students and two Capital students produced the same objects that Western students would produce, i.e., part of a cube for (a) and a triangular prism for (b). Perhaps the best way to indicate the other students' problem is to say that if (a) *is* part of a cube then the plan view should look like part of a square. However, if the plan view *is* that then the front view will not be as shown in Figure 8.1(a).

Several students made shapes which had the 'correct' front view, i.e., as on the card but the plan view was either (a) or (b) in Figure 8.2.

There is no information in the diagram which says how long the side labelled "x" is; in Western cultures decisions on such matters are made on the basis of visual experience with foreshortening and practice with the oblique convention.

Other tasks, which not only involve understanding conventions but also the application of other skills were even harder. It has often been reported that students from non-Western cultures are poor at spatial skills, but it is often forgotten that 'pictorial' spatial tests invariably involve conventions. We are so familiar with these that we take their knowledge for granted and assume a universality of understanding which is quite erroneous.

Conventions are of course learnt, as are the reasons for needing them, *and* the relationship between the pictures and the reality that are conventionalising. The hypothesis is therefore provoked: perhaps much of the found difficulty with spatial tasks lies in understanding their conventions, and that if these *are* known by those people, from both non-Western and Western cultures, who are supposedly weak spatially then perhaps they would not appear to be quite so incapable.

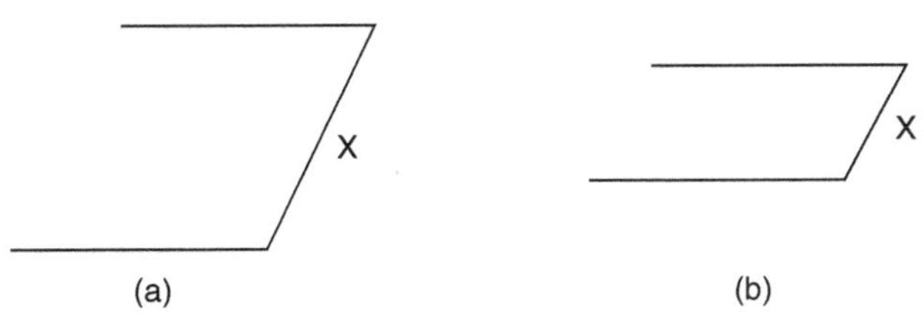

Fig. 8.2.

Drawing

Several of the tasks required the students to draw and three tasks in particular are illustrative for our purposes here. In one task, the students were asked to copy the drawings from a specimen set, produced from Plate 1 of Bender (1938). The drawings use straight and curved lines, dots, closed and open shapes, geometric and irregular shapes.

This task revealed two types of difficulty. First the obvious lack of expertise at drawing and copying. Much erasing, head scratching and tongue-clicking was in evidence particularly after the student had drawn something and was then comparing his effort with the original.

The other difficulty with this task (and with others) was the criteria to be satisfied. Once again "copy" implies to us "identical". But "How accurate is accurate?" seemed to be their unasked question. So, scales varied, lines bent, angles varied, and curvatures altered. Of course, if Westerners attempted to draw and copy PNG patterns and designs, they would often make similar 'mistakes' through ignorance of the criteria to be met. There is nothing obvious or logical about criteria like these. They must be learnt. In another task, each student was presented with a small wooden block made from 1 cm wooden cubes. 19 cubes were used and the student viewed the block from across the table – his view is shown in Fig. 8.3. The student was to sketch the block as it appeared to him.

In this task, unlike the previous one, the student must decide what to include and what to omit, and he must imagine the 'ideal' picture that he is trying to reproduce. Several students could not remember ever having been taught how to draw real objects. It was possible to obtain improved drawings by pointing out specific clues like "keep verticals vertical" and "keep parallel lines in the object parallel in the

Fig. 8.3.

drawing". (These are both adequate hints for drawing small objects.) Advising the student to close one eye helped also, emphasizing that *we* take the photographer's one-eyed view of the world very much for granted. We don't realise how much it conditions us in our drawings.

Map drawing, whilst by no means appearing simple, seemed to be a more familiar task. The students could have learned this at school or perhaps they find this a more natural and sensible use of visual representation than the drawing of objects from strange angles. Later, when I asked them about their village gardens or fishing areas some of them enthusiastically drew me sketch maps with many details included. I found that the maps the students drew were in the main adequate for the communication purposes they were meant to serve. They were as accurate as they needed to be.

One intriguing finding was that when two of the students were asked to draw a map of the campus showing their route from their room to my office, they produced maps which contained no roads, only buildings. Both were born in the island region where roads, as we know them, were non-existent.

These tasks, then, point to some of the skills of drawing, and to the criteria to be satisfied, particularly to the recognition of the purpose to be fulfilled by the drawing, by which accuracy is judged. This seems to me to be one of the most important values of drawing – that by doing it one learns *about* drawing and one is enabled to read other people's drawings.

You can only read this text because you know the conventions employed. In schools reading and writing are usually taught concurrently and the whole complex procedure of forming the letters, writing words, keeping to the line, writing from left to right, leaving certain spaces, etc., is learned by having to be a "user" of conventions, by being a writer, not just a reader.

Visualising

This ability is, for me, right at the heart of any spatial work, and I was interested to see the quality of visualising in the students I was working with. Reports of other research (e.g., Philp and Kelly, 1974) suggested that 'ikonic processing' was likely to be the predominantly used cognitive strategy. Other studies (Lean, 1975) suggested that students were weak spatially, largely on the basis of group spatial testing. My first impressions were toward the latter view, but as the work progressed I started to think that if the object was well known, and the convention used in representing it was a familiar one, then imagining and visualising with regard to that representation would be well done.

One task which showed this was the sub-test "Matchbox Corners" from Spatial Test 2, produced by the National Foundation for Educational Research. A matchbox was drawn with dotted "hidden lines" and a black dot placed on one corner. Four drawings of the matchbox, rotated in 3D space, were then presented and the student was asked to draw a black dot on the corresponding "same" corner of each. Five different sets were used and the task was presented here untimed.

Very few errors were made and yet the task is known to involve a high degree of visualising ability.

Another task which illustrated their strength was the sub-test "Word Recognition" from the Multi-Aptitude Test, Psychological Corporation, U.S.A. 18 typed English words were presented in varying degrees of obliteration and the student was asked to write the original word. Despite the fact that English was each student's second, third or fourth language, they did remarkably well at this task. By contrast, a line diagram counterpart of the previous task was very difficult for the student. It was produced using drawings from Kennedy and Ross (1975). The drawings showed the outlines of 'familiar' (to these students) objects, e.g., people, animals, birds, aeroplanes, house, car. Two forms were presented, one with approximately 80% omitted, the other with approximately 40% omitted. The student was asked what the diagrams showed originally.

The behaviour change from the previous task was fascinating to watch. Whereas for the word-completion the students often "drew" letters with their fingers to help them imagine the word, they did not do this with the incomplete line drawings. They merely looked, occasionally turned the paper round and guessed very hesitantly. Clearly, even if the "objects" were known to them the representations of them were not. Again, the contrast with the word-completion was marked – they had been taught the written representation of English words for several years at school, but not the drawings.

Finally in this section a task which illustrates the strong link (for these students, at least) between visual *memory* and visualising.

12 small everyday objects (e.g., coin, key, pin, etc.) were set out on a 3 × 4 rectangular board. The student was given 45 seconds to look at the arrangement, the objects were then tipped off the board and the student was asked to replace them correctly. Only one student made any error. He had only two adjacent objects wrong, and was suffering from malaria at the time!

There was concerned attention given to this task by all the students, who in most cases replaced the materials carefully and deliberately. The typical Westerner attempts this quickly, as if he were trying to get the right answer before his memory faded, and it was, therefore, interesting to see how long the memory stayed with these students. Certain students were presented with the objects again a day later and were sucessful at replacing them, a week later (one student 10 out of 12) and the same student two weeks after the initial viewing (all correct – he had, therefore, corrected his mistakes of a week previously!) This delayed request was clearly not unreasonable, and most of the students attempted the task as if they were confident of success.

Other stimuli were used, with varying degrees of success. Feathers and playing cards were the two most difficult stimuli, and it was clear that no student was using a verbal coding. "I just remember how it looks" was a typical comment. Even some of the Islander students who knew some of the shells by name did not name them for this task. They were quite surprised when I suggested that some people might remember the location of a shell by using its name. The colour, the shape, the texture and the size, all were used, but not the name. As one student pointed out, some of the shells had the same name so that wouldn't help – the fact that they almost looked the same didn't worry him!

This last point is important, and supports the reports of other researches concerning "ikonic processing". In several of the tasks no verbal mediation was evident from the students even though they could have used it. Very little was said at all in fact, unless it was in answer to a question or because the task sought an oral response. Much looking (pointedly), head turning, paper turning, and moving backwards and forwards (as if to alter the focus) was in evidence – all suggestive of a "behavioural support system" for visual strategies. No words though.

Language

The problems caused by local languages which are not designed for mathematical and scientific use are becoming increasingly well known. One task which will illustrate part of the difficulty is this one. Individually the students were asked to translate a list of 70 English words into their own local languages. Interestingly enough, only the following words were able to be translated by all twelve students: below, far, near, in front, behind, between, middle, last, deep, tall, long, short, inside, outside and hill. Some of the words that were omitted (i.e., difficult to translate, or forgotten) by more than half the students were: opposite, forwards, line, round, smooth, steep, surface, size, shape, picture, pattern, slope, direction, horizontal and vertical.

From the Westerners' mathematical point of view, then, there were gaps in language, and equally there were many overlaps where the same local word is used as the translation for several English words. One example was given by a Manus student who reported that each of the following words translates into the same word in his language: above, surface, top, over and up. Although there were a few overlaps in each language, most of them occurred with the Manus languages. Many confusions could easily occur in school mathematics and science because of the need to distinguish, for example, 'side' from 'edge' which couldn't be done easily in one of the Enga dialects, nor in one of the Manus languages.

Another interesting point was that 'above, nearest, forwards and first' were omitted more often than their partners 'below, furthest, backwards and last'. This suggests that sometimes the 'negative' term in a pair of polarized comparatives is more often used than the 'positive' term, a result which would appear to be in conflict with the findings of linguists (e.g., Clarke, E., 1972; Donaldson and Wales, 1970).

Papua New Guinea must be a linguists' paradise as there are thought to be about 750 different languages spoken there, and several of them can now be written. There is some fascinating research which is developing on the relationship between language, classification systems and counting systems. Lancy (1977), for example, has data from several sources on 150 different counting systems which he is currently analysing in terms of classification – ask yourself, how many different counting systems do you know? Many of the languages appear to have no conditional mood – you cannot easily say "If . . . then". This then provokes the question: if it is not said, is it ever thought? Classification does not appear to be hierarchical as for us, e.g., there can exist words for specific shapes but no word for 'shape', and as Kelly and Philp (1975) say "even where the language is perfectly adequate to form a hierarchy, the children *do not*, in fact, do this as a matter of course" (p. 194).

Another researcher (Jones, 1974) asked local interpreters to try to translate some mathematics tests into the local language. Many questions were impossible or very difficult to translate. Some examples of the replies were:

"There is no *comparative* construction. You cannot say *A* runs faster than *B*. Only, *A* runs fast, *B* runs slow."

"The local unit of distance is a day's travel, which is not very precise."

"It could be said (that two gardens are equal in area) but it would always be debated."

For comparing the volume of rock with an equal volume of water, "This kind of comparison doesn't exist, there being no reason for it", and hence you cannot say it!

It is not, of course, merely a matter of teaching the language, because spoken language is only an observable result of some unobservable thinking. Differences in language imply differences in thought. So, if you ask 'deeper' questions as I did of a local anthropologist a different order of difference becomes recognisable. As she said in a personal communication to me (Biersack, 1978):

"Paiela (a Highland group) space has some unique properties:

(i) It is not a container whose contents are objects. It is a dimension or quality of the objects themselves, as their locus.

(ii) Space is a system of points or coordinates as the loci of objects. Objects are defined through binary opposition, as large or small, long or short, light-coloured or dark-coloured; and space, as the coordinates of objects so defined, becomes axial rather than three-dimensional, as up or down, over there or here, far or near, and so on.

(iii) Space is not objective but the product of the observer's perception of opposition in sensory data.

Among other things it means that size (for them) would be like value (for us), not absolute or gauged by objective measures but relative, dependent upon the subjective factors of evaluation and scale of comparison."

Value is seen in comparison. Hence she says of pig-exchanges,

"so long as the actual pig has not yet been produced, it is impossible to know its size. Once the pig is actually given, and once it is actually placed in proximity to other pigs it is possible to evaluate it large or small... The uncompared pig is attributeless or 'unknown' while the compared pig has at least one attribute that can be 'known'."

With another group, the Kamano-Kafe, in the Eastern Highlands the four 'units' of length are 'long', 'like-long', 'like-short', 'short'. Similar *adjectival* rather than *invariant* units are also reported from other areas (Jones, 1974).

So, Western conceptions of space with its ideas of objective measurement are not universal, nor are they 'natural', 'obvious', or 'intuitive'. They are shaped by the culture. They are taught, they are learnt.

Cognitive Characteristics

I have referred in this article mainly to spatial ideas, because that was the focus of my research, and that is where my main interest lies. But in reading reports of other research carried out in Papua New Guinea, in talking with other researchers, and in working with the students at the University of Technology I became increasingly aware of several differences in what I call 'cognitive characteristics' between PNG students and the students I work with in the U.K.

The most striking point was their concern with the specific as opposed to the general. Their languages seem to have many specific terms, few general ones. The classifications and taxonomies used in their culture seem to have few hierarchies. Generalising is not the obvious mode of operating there as it appears to be for us – there not only seems to be a difficulty with doing it, there is felt to be no need to do it. Indeed I sometimes had the feeling that I was rather crazy when I tried to operate in a generalised and hypothetical way.

For example, I asked a student "How do you find the area of this (rectangular) piece of paper?" "Multiply the length by the width". "You have gardens in your village. How do your people judge the area of their gardens?" "By *adding* the length and width". "Is that difficult to understand?" "No, at home I add, at school I multiply". "But they both refer to area". "Yes, but one is about the area of a piece of paper and the other is about a garden". So I drew two (rectangular) gardens on the paper, one bigger than the other. "If these were two gardens which would you rather have?" "It depends on many things, I cannot say. The soil, the shade . . . " I was then about to ask the next question "Yes, but if they had the *same* soil, shade . . . " when I realised how silly that would sound in that context.

Clearly his concern was with the two problems: size of gardens, which was a problem embedded in one context rich in tradition, folk-lore and the skills of survival. The other problem, area of rectangular pieces of paper was embedded in a totally different context. How crazy I must be to consider them as the same problem!

As Biersack (1978) again said: "With regard to the ability to generalise, I think *on principle* the Paiela do not generalise. They have, rather, a problem-solving approach to everything. Every problem is a unique set of circumstances having a unique solution, and you cannot solve problems in the abstract, you can only solve them within the context of the particulars of the problem. I don't think this approach excludes an appreciation of general principles. I said it was *on principle* that the approach was adopted. It's just that the principles of 'their' approach and 'our' approach are different."

When this type of thinking operates it seems that many of the teaching strategies which I know about become meaningless. The use of analogy, the use of counter-examples, strategies which are designed to foster understanding, or discovering general principles. All of these assume the acceptance of generalising, hypothetical thinking, and hierarchical processing, as important and worthwhile ways to behave.

Conclusion

Earlier in this article I said that I was not going to discuss strategies for educational development in Papua New Guinea, although as you can probably infer, I find that problem both fascinating and formidable. My concern here was to offer you some data which, I hoped would contrast in various ways with the data you would normally meet.

But how successful have I been? Leaving aside those readers who work in Papua New Guinea or similar cultures, consider how different this data is. Even in technologically developed societies and cultures do we not find some of these problems – sometimes with adults, certainly with children? Could I give as a general description of those people "those who have not yet been inducted into the mathematician's culture" or sometimes even "those who have chosen not to enter it"?

Perhaps if we consider mathematics education as a form of cultural induction we would realise both the enormity of the task and the range of influences that can be brought to bear. We would, for example, not only consider problems like "What are the skills necessary to be a successful mathematician?" but also others like "What is the value of entering the mathematician's world?" and "Why do we consider it to be so important?" If we do consider mathematics to be problem-solving par excellence, then we should also recall that it is only *one* approach to problem-solving and it can be seen by 'outsiders' as a very strange business. (As another example, ask yourself why you spend a long time looking for a quick solution.)

Even if we feel we know what the values of learning mathematics are we then face problems such as how do we transmit those values? What do we know about the role of the teacher as a cultural transmitter, as an example, as a model for imitation? And there are many other questions.

Mathematics education has powerful cultural and social components. Perhaps we should give them the attention which we have already given to the psychological components.

University of Cambridge, England

References

Bender, L., *A Visual Motor Gestalt Test and its Clinical Use*, American Orthopsychiatric Association, New York, 1938.

Biersack, A., Personal communication from Dept. of Anthropology, University of Michigan, U.S.A., 1978.

Bishop, A.J., *Visual Mathematics*, Proceedings of ICMI/IDM Regional Conference on the Teaching of Geometry, Universität Bielefeld, West Germany, 1974.
Bishop, A.J., *On Developing Spatial Abilities*, A report to the Mathematics Education Centre, University of Technology, Lae, Papua New Guinea, 1977.
Bishop, A.J., *Spatial Abilities in a Papua New Guinea Context*, Report No. 2. Mathematics Education Centre, University of Technology, Lae, Papua New Guinea, 1978.
Clarke, E., 'On the Child's Acquisition of Antonyms in Two Semantic Fields', *Journal of Verbal Learning and Verbal Behaviour* **11** (1972), 750–758.
Deregowski, J.B., Teaching African Children Pictorial Depth Perception: in Search of a Method', *Perception* **3** (1974), 309–312.
Donaldson, M. and Wales, R. J., 'On the Acquisition of Some Relational Terms', in J. R. Hayes (ed.), *Cognition and the Development of Language*, New York, Wiley, 1970.
Jones, J., *Quantitative Concepts, Vernacular and Education in Papua New Guinea*, Educational Research Unit Report 12, University of Papua New Guinea, 1974.
Kelly, M. and Philp, H., 'Vernacular Test Instructions in Relation to Cognitive Task Behaviour Among Highland Children of Papua New Guinea', *British Journal of Educational Psychology* **45** (1975), 189–197.
Kennedy, J. M. and Ross, A. S., 'Outline Picture Perception by the Songe of Papua', *Perception* **4** (1975), 391.
Lancy, D. F., *The Indigenous Mathematics Project: A Progress Report*, from the Principal Research Officer, Department of Education, Konedobu, Papua New Guinea, 1977.
Lean, G., *An Investigation of Spatial Ability Among Papua New Guinea Students*, in Progress Report 1975, Mathematics Learning Project, University of Technology, Lae, Papua New Guinea, 1975.
Philp, H. and Kelly, M., 'Product and Process in Cognitive Development: Some Comparative Data on the Performance of School Age Children in Different Cultures', *British Journal of Educational Psychology*, **45** (1974), 248.
Strathern, A., 'Mathematics in the Moka', *Papua New Guinea Journal of Education* **13** (1977), 16.

Chapter 9
Cultural and Social Aspects of Mathematics Education: Responding to Bishop's Challenge

Bill Barton

Bishop's (1979) paper, *Visualising and mathematics in a pre-technological culture*, brought to our attention cultural difference in mathematical visualisation not just as an anthropological observation (cf. Deregowski, 1972, 1974, 1984, on visualisation in Africa), but as a challenge to mathematics educators to reconsider the nature of their task. He illustrated cultural difference in mathematical skills that are valued (the ability to visualise objects in space), in mathematical conventions and criteria (what real world problems are "sensible"), in mathematical habits, in taught experiences, and in the structure of language used in mathematical thinking (in particular expressing conditionals). By noting that all of these are culturally based, and far from universal, he challenged us to "consider mathematics education as a form of cultural induction" (p. 145). Such a view would, he suggested, alert us to both the enormity of the task and the range of influences on the result. What responses has the mathematics education community made to this challenge?

Bishop himself subsequently elaborated this theme in his (1988) book Mathematical Enculturation. Such a substantive challenge to the community was always going to generate other responses: creative, multidisciplinary, scholarly, or unexpected.

Responses to the Challenge

In the language arena, these issues had already been raised. The 1974 Nairobi UNESCO Symposium on the Interactions Between Linguistics and Mathematical Education (UNESCO, 1975), was a precursor to a wider consideration of the cultural aspects of mathematics education. But, at that time, the issue was one of finding ways to improve the learning of school mathematics. Bishop's insight was to see that mathematics education could be viewed as a cultural act as validly as it could be seen as acquisition of knowledge. The literature on language and mathematics is now extensive, and regularly ventures into political, philosophical, or

B. Barton
Dept. of Mathematics, The University of Auckland, Private Bag 92019, Auckland Mail Centre, Auckland 1142, New Zealand
e-email: b.barton@auckland.ac.nz

P. Clarkson, N. Presmeg (eds.), *Critical Issues in Mathematics Education*,

psychological territory (see, for example the recent Special Issue of Educational Studies in Mathematics (Barwell, Setati, & Barton, 2007)).

Without suggesting a direct response to Bishop's paper, the cultural induction challenge articulated there has been taken up elsewhere within a multidisciplinary context in several ways. Sociological approaches to mathematics and mathematics education have built on from the work of Spengler (1926, 1956) with the writings of Bloor (1976, 1994) and Restivo (1983, 1992, 1993); historical accounts of mathematics have been extended to arguments of historiography (see Daubin, 1984, 1992; Crowe, 1975, 1992; Wilder, 1950, 1968, 1981); and anthropologically mathematics has been addressed in the work of, for example, Pinxten, van Dooren, and Soberon (1987) or Senft (1997).

A further scholarly response to the need to introduce a cultural perspective into our thinking about mathematics and mathematics classrooms, has been the more post-modern investigations of people such as Dowling (1998) using semiotic analysis of mathematics texts; Davis (1996) using enactivist thought and critical theory; and Brown (2001) using hermeneutics, all of whom use the work of Bourdieu, Foucault, Bakhtin, and Luria.

A response I found unexpected, although I should not have done so if I had reflected for a moment, was the resistance to any idea of mathematics as culturally based. Acknowledging the cultural elements of mathematics education has been treated as the thin end of the wedge that is prising open the oyster of mathematics, and letting subjectivity or relativity, creep in. The Math Wars in America broke out over terms like "constructivism" and "basics" (codes for "relativity" and "objectivity"), but need to be seen in a much wider context. Joseph Fiedler, in his endorsement of Latterall's (2004) Math Wars: A Guide for Parents and Teachers, describes the book as an explanation of "how Mathematics (of all things) has become a battleground in the culture wars that characterize our society".

Thus the cultural awareness of our age, addressed by Bishop over several decades, has had many expressions. However, in this chapter, I wish to focus my attention on the creative response of the mathematics education community to the direct cultural challenge Bishop articulated. A whole new field has arisen that concerns itself with culturally specific mathematics and its role in mathematics education: the field of ethnomathematics.

Ethnomathematics and Mathematics

Ubiratan D'Ambrosio must have been aware of Bishop's specific challenge when he gave his 1984 ICME plenary session "Socio-Cultural Bases for Mathematical Education" during which many heard the word ethnomathematics for the first time (D'Ambrosio, 1986). How well has the subsequent development of this field met Bishop's challenge?

First let it be noted that in discussions with Bishop he has said to me that he does not regard his work as ethnomathematics. Nevertheless, his plenary at PME in

Noordwijkerhout in 1985 was instrumental in bringing cultural issues to that community. Bishop has, rather, pursued the issue of cultural conflict within the mathematics classroom. He came to this by first shifting his research orientation from mathematics teaching to examining the way mathematical meaning is socially constructed by students (Bishop, 1985). This gave him three foci of attention: activity, communication and negotiation. It is not surprising that he moved into the inevitable conflicts that arise because meaning is culturally-based and therefore an individual's constructed meanings will be at odds with those of others from different cultures (Bishop, 1994). This being especially significant when the culture of the teacher differs from that of a student or group of students. These directions of Bishop's work are taken up elsewhere in this volume (see, e.g., chapter 17, p. 237 ff.).

Second, although D'Ambrosio was, in 1984, definitely aware of Bishop's writing, his motivation for ethnomathematics emerged not only from his Brazilian experiences, but also from a period in West Africa and his observations of the teaching of students from different backgrounds in America.

Be that as it may, the field of ethnomathematics, or the "research program on the history and philosophy of mathematics with pedagogical implications" (D'Ambrosio, 1992), has grown dramatically in 20 years (for example, the three international conferences on ethnomathematics in 1998, 2002, and 2006 (Contreras, Morales, & Ramírez, 1999; Monteiro, 2002; Barton, Domite, & Poisard, 2006)). And it has addressed Bishop's initial challenge. If mathematics education is a form of cultural induction, then what is it an induction to, how does it interact with other cultural modes, who decides the subject of the induction, and how could the induction proceed?

The first of these four questions is a question about the cultural nature of mathematics. The second is about the relationship between mathematics and other activities. The third is about the politics of mathematics education, and the fourth is about how to include culture in the mathematics classroom. In summary, we can say that ethnomathematics has made progress on these challenges more slowly than was initially expected, but that it has developed a sound theoretical base from which to answer these four questions.

The distinct mathematicality of many cultural practices is now well established through a plethora of studies. The reader is referred to the proceedings of the three conferences to grasp the scope of cultural activities studied. Many such studies describe cultural activities in the terms of standard mathematics (what I refer to as near-universal, conventional mathematics or NUC-mathematics; what Bishop distinguished by talking about Mathematics rather than the more generalised mathematics).

But it is not much to say that we can describe culturally specific activities using NUC-mathematics. Ethnomathematics has progressed beyond that. There are several examples where the mathematics that is described therein has a cultural distinctness. Examples include the relationship patterns of Australian aboriginal genealogies (Cooke, 1990), the categorisation of weaving patterns by weavers (Barton, 1996), and distinct calculation techniques (Albertí, 2007). More than that, there are at least three cases where new mathematics has arisen from ethnomathematical considerations: the visual computer language embedded in Indian Kolam drawings (Ascher, 2002); the curved "linear" functions of double-origin geometry inspired by ways of reference in Polynesian languages (Barton, 2008, pp. 22–24); and, most recently, the cyclic matrices inspired by Angolan sand-drawings (Gerdes, 2007).

It may be argued that these are modern manifestations of the way that mathematics has developed throughout history. Much mathematics has emerged from geographically or culturally distinct sources, that is what mathematics is. NUC-mathematics is an amalgam of abstractions and generalisations about the spatial, relational, or quantitative aspects of specific practices that have been generalised and abstracted into sophisticated systems. What such a description misses is that the way such an amalgam is integrated, the language and symbolisms used to describe it, and the aspects of the systems that are made dominant or that get more attention in the development of the field, are all the result of cultural preferences. In the history of mathematics, the last 500 years of such development has been dominated by European, Russian and American philosophies, languages, thinking, and political or social imperatives. Modes of thought and orientations familiar to indigenous societies, Eastern cultures, or by more artistic or artisan communities have been ignored, or, at best, subsumed (for a discussion of how this occurs, see Barton, 2008, pp. 108–115).

The second question is often implied by critiquers of ethnomathematics. Surely, they argue, street vendor calculation is money exchange, not mathematics, weaving is weaving (see, for example, Rowlands & Carson, 2002, or Horsthemeke & Schäfer, 2006). We should not be trying to turn these into mathematics, however much we might use them as examples of alternative calculations or spatial patterns. Mathematics, they imply, stands above these practices as an abstract system that has multiple exemplifications. The exemplification is not the system.

What, then, does ethnomathematics have to say about the relationship between mathematics and otherwise described practices such as tailoring, building, recreational games?

One way of answering this question is to think of the relationship in the same way we might think of the relationship between pure mathematics and applied mathematics, and between them and applications of mathematics. Modern disciplines like economics could not exist without sophisticated mathematical techniques, and, indeed, economics has given rise to original mathematics. This is not to claim that one is the other, but merely to note that we can look at some of the activities of economics from a mathematical point of view, and some of pure mathematics as sourced in, and closely related to, economics.

Another way to consider the question is to ask what it is that mathematics does. Amongst other things, it is the way we understand quantitative, spatial and relational aspects of our world – it is the language we use to speak of these things and understand them better. Under such a definition, any system that achieves this outcome might be legitimately regarded as mathematics, whether it is found in a school mathematics textbook or in an artisan's language and hands.

However, the field of ethnomathematics has moved beyond this territorial type of discussion that is likely to end up in futile land wars, and begun to focus on process. Rather than seeking the ossified mathematics in a practice by simply describing the system, ethnomathematicians are now moving in two directions. One has already been noted above, and is best exemplified in the work of Gerdes. Since D'Ambrosio's talk he has been producing mathematics himself through contemplation of cultural artefacts (Gerdes, 1986). The matrices mentioned above are a mathematically significant product of this endeavour. Ascher (1991) did similar work with a variety of cultural activities. This is not to claim that such mathematics is in these practices or activities, but that the cultural work inspired such mathematical thinking.

Apart from generating new mathematical thinking, ethnomathematics has promoted authentic dialogue between artisans of various kinds and mathematicians. The process is best exemplified in the work of Alangui (in process). As part of his work investigating the mathematics of the rice terraces in the Philippines, Alangui became the intermediary in a dialogue between the rice terracers and university mathematicians. Focussing on how water flows are controlled, mathematical models were set up on the one hand, and practical experience plus an in depth knowledge of cultural practice was contributed on the other. There is potential for a new kind of mathematical activity in such dialogue, activity that both reflects in a modern mathematical way on a cultural practice, but at the same time keeps the presentation of the cultural practice with those to whom it belongs. Rather than the practice being seen through mathematical eyes (and potentially being devalued in the process by invidious comparisons), the cultural activity, and the concepts that underlie it, are used to critique the mathematical interpretation in a two way process of some equality.

Ethnomathematics and Mathematics Education

The third and fourth questions that arise from Bishop's challenge are both questions about mathematics education and ethnomathematics. If there are cultural forms of mathematics, who decides which form of mathematics becomes the subject of schooling, and how can culturally specific forms be properly presented in the classroom? In my view, the answers to both these questions remain very open.

Mathematics education has always been highly political. The American "Math Wars" have already been cited and debates about school curriculum reaches the very highest political levels (see, for example, Ellerton & Clements, 1994; Loveless, 2001). Bishop radically addressed these issues in his provocative paper in *Race & Class* (Bishop, 1990). What has the writing on ethnomathematics to say about who should determine curricula? (For a wider investigation of the political question, please see chapter 12 by Keitel and Vithal in this volume, p. 165 ff.).

D'Ambrosio's writing has always spoken of the hegemony of NUC-mathematics in mathematics education, and the need for new conceptualisations of curriculum

(D'Ambrosio 1980, 1990, 2001). Has the ethnomathematics community responded? Yes, the answer is unequivocal: those being taught. Two main writers have developed this response in theoretical ways: Frankenstein and Knijnik. What is interesting is the different approaches used.

Frankenstein began with critical theory. Her 1983 paper takes Freire's critical theory and directs its light towards urban mathematics education. Her subsequent writing, often jointly with Powell, has developed both methods and approaches that respond to the needs of urban adult learners in America, particularly non-white Americans (Frankenstein, 1998; Powell, 2002; Powell & Frankenstein, 2006).

Knijnik, on the other hand, has developed theory out of practice. Her work with the Landless Peoples of Brazil has continued over many years (see, for example, Knijnik, 1993; Knijnik, Wanderer, & de Oliveira, 2005), and although she has drawn on theorists like Foucault, her writing "aims at discussing the conditions of possibility (the social, political and cultural context) for the emergence of statements about adult Brazilian peasant mathematics education, how such statements circulate in peasant pedagogical culture, and their effects of truth on school mathematics processes." (Knijnik, 2008, abstract). In other words, Knijnik has given us a way of looking at cultural perspectives on school mathematics in those situations where the community receiving the education has little to do with the curriculum.

The result of this writing is more to critique the hegemony of most school mathematics curricula, rather than to suggest ways this can change in anything other than a very local context. In other words, Bishop's challenge remains as a curricular issue, but we now understand more about the mechanism by which curricula are resistant to cultural input.

Fortunately we have made more progress on the question of how, in practical terms, cultural material may be introduced into a mathematics classroom in an authentic and effective way. From the very early attempts (e.g., Zaslavsky, 1991a, 1991b), where motivation and inclusiveness were the explicit aims of cultural input, where much activity took place in pre-service teacher education programmes, and where research in the area amounted to recording the (nearly always positive) affective responses from students and teachers, we now have a theoretical and research base of some sophistication.

The work of Jerry Lipka and his team at University of Alaska Fairbanks has led the way in classical comparative research of the effects of a well-developed community-based cultural intervention in mathematics (Lipka et al., 2005; Lipka & Adam, 2006; Lipka, Sharp, Adams, & Sharp, 2007). Not only are the interventions the result of several years of working with the community, but also there have been controlled studies of the mathematical learning effects on standard tests with students exposed to the programme. The evidence available shows that such programmes are beneficial, although the researchers are understandably cautious about generalizations. Other examples of such work exist (e.g. Adam, 2004; Hirsch-Dubin, 2006), but are not as comprehensive.

There have been critiques of cultural material being presented in mathematics classes (e.g. Horsthemeke & Schäfer, 2006; Rowlands & Carson, 2002; Vithal & Skovsmose, 1997). Acknowledged dangers include the risk of misrepresenting or devaluing cultural material by taking it out of context, further marginalising cultural minority students if they are not familiar with their "own" cultural material, and presenting mathematically unsophisticated material as more significant than it really is.

Most of thc powcr of these critiques is diminished in situations like that with the Yu'pik programme in Alaska where there is strong involvement of both mathematics educators and cultural elders, and support from both community and curriculum authorities. Further strength is given to more recent interventions by the developing theorising of the work. For example, Adam (2004) describes five different models of an ethnomathematical curriculum present in previous work, and then elaborates her own model that positions ethnomathematical material as a bridge that helps students understand the nature and key ideas in mathematics from the understanding of mathematical aspects of their own culture. Such a model is effective primarily in situations where the class is predominantly from one culture that is not close to the culture of mathematics.

An unanswered aspect of the possibilities for an ethnomathematical curriculum is the larger question of the impact on curriculum itself. Is it possible (let alone useful or effective) that the nature of curriculum can be made more culturally specific through an ethnomathematical approach? To be specific, for example, can (and should) mathematics classrooms be effective places for students to learn mathematical modes of thinking that are different from those associated with NUC-mathematics? What is the role for such education? To highlight the difference between this question and the interventions described above, consider the issue of researching such classrooms. Comparative controlled testing against standard mathematics tests would not be appropriate. The outcomes of a "true" ethnomathematical curriculum would need to be assessed against the stated objectives of such a curriculum, of which standardized test performance would be a minor part, if it had any part at all.

We may summarise by saying that the fourth question implicit in Bishop's challenge is being answered by ethnomathematics. We now know some ways to make mathematics a more effective "form of cultural induction" by using cultural materials and approaches in some particular situations. We are also more aware of the difficulties, and we have a clear vision of where we need to move in the future, both theoretically and with respect to research. In particular, we want to know how far the empirical evidence can be generalized across other types of programmes. However, the weight of evidence for ethnomathematical approaches in the classroom is positive.

To conclude this first part of the paper, there is no doubt that Bishop's challenge to the mathematics education community to work through the consequences of regarding mathematics education as a form of cultural induction has been taken up by the ethnomathematics community in many forms and progress has been significant, if slower than many hoped. The field is expanding, both in its multiple agendas, but also in the number and type of people wishing to become involved. General educationalists are looking to the mathematics example for inspiration, mathematicians are turning their attention to cultural issues, and teachers in all sorts of classes are experimenting with ethnomathematical materials, sharing them, and debating their effects.

The Contingency of Mathematics and Mathematics Education

Bishop's 1979 paper raised the contingency of many aspects of a person's mathematical experience. Bishop's insight into our experience of mathematics is to say that he questioned the extent to which mathematics is unique in its nature and content. What if mathematics is as contingent, socially constructed, and culturally differentiated as is, say, history, architecture, or music? How far have mathematics educators progressed along this path? What more do we know, 25-plus years on, about the socio-cultural contingency of mathematics and mathematics education?

We know a lot more. But more importantly, we have moved forward significantly in the way we know about mathematical contingency. No longer is mathematics beyond cultural or social analysis because no longer is it accepted that the subject deals with cultural universals and a priori knowledge. Mathematics is now widely accepted as a human construction at some level. In order to achieve this move into more social and anthropological points of view as relevant for mathematics, it has become necessary for mathematics educators to understand, and use, the theoretical constructs of other fields. Three examples are briefly discussed.

Bishop's 1988 book was one of the first to address the important construct of culture, and acknowledge that it has moved over time. Alangui (in process) contains a more detailed analysis in a mathematical context, but readers are referred to anthropology writers to fully understand the issues (e.g., Ray, 2001). Certainly, older conceptions led to issues of alterity, that is, defining cultures as "the other", and hence making them different and exotic. Modern conceptions involve a notion of constant change, and cultures are defined in interactions with each other. The implications of this stance for a view of mathematics as socially and culturally constructed, are that it is very difficult to create a stable relationship between any form of mathematics and any culture, certainly any individual. Rather, the idea of cultural construction of knowledge is more nuanced, more time dependent, more integrated with other social and individual factors than early writing suggested. An example of this aspect is Skovsmose's ideas of Foreground and Background when considering students' mathematical responses (Skovsmose, 2002). Another consequence is a more equal consideration of different cultures or social backgrounds.

Another area in which progress has been made are investigations into mathematical cognition. With respect to cultural and social influences, the idea of embodiment of mathematics has led to a renewed understanding of our ability to make mathematical knowledge part of ourselves, to integrate it with our actions and symbols (see, for example, Tall, 2006). As humans we are agents of change, we do things, we make things happen. An awareness of how, for example, in algebra, we might use our experience of balance to understand solutions of equations, alerts us both to the power of embodiment, and its limitations. If students use their concept of balance in the real world to inform their mathematical constructions, then all the different features of different students' real experience is likely to be subtly played out in their mathematics. Cognitive theorists often take all humans as the same, but to the extent that writers such as Lakoff and Johnson (1980, 1999) are correct, we have a way of analyzing some cultural difference.

From linguistics the concepts of discourse and grammar have added to our understanding of mathematical knowledge. Halliday (1975), in a paper prepared for the 1974 UNESCO conference in Nigeria first raised the idea of mathematical discourse. This construct has now emerged as a key tool in understanding the interactions between language and mathematics in both monolingual and multilingual situations. Halliday's original theories have been developed (O'Halloran, 2005), and discourse is now one tool in our armoury to understand the development of mathematical thinking (Hersh & Umland, 2005). Again, to the extent that discourse in general, and mathematical discourse in particular, differ in different languages, we have access to cultural difference in children's mathematical understanding. With respect to Mandarin-speaking students, the mathematical characteristics of discourse have already been well-documented (see Galligan, 2001, for an overview). For implications in multilingual classrooms, see a review by Setati and Adler (2001) and also Barwell, Setati, and Barton (2007).

Post-Modern Tentacles

Mathematics may be one of the last spheres of human knowledge to be challenged as a stable and static edifice and to be seen, like all human knowledge, as subject to time, context, and human proclivity. We can now see Bishop's words as the tentacles of post-modernism wrapping themselves around the final masthead of firm knowledge. Has the good ship Mathematics been dragged under the waves by this monster, or are the resisting sailors (also known as mathematicians) keeping her afloat above the wishy-washy sea?

In true post-modern style, of course, the ship, the monster, and the sea have redefined themselves with respect to each other, thanks in part to Bishop's continuing work over the years to use interdisciplinary concepts with respect to mathematics. The ship is a member of a flotilla of vessels that share both sailors and equipment, and support each other. That is, mathematicians are not a tightly defined group, but occur in many different spheres of knowledge doing many different sorts of things. The subject mathematics is more clearly now seen as intertwined with other knowledge. This point is also more (but not completely) accepted within the community of mathematicians.

The monster is not necessarily trying to drag the ship under the waves, but, rather, is using the ship to draw itself out of the sea and ride with the ship as it travels, even pushing it along. That is, considerations of language, hermeneutics, anthropology and other socio-cultural areas enhance mathematics, they do not destroy it. The more we learn about the nature of mathematical knowledge, the more we are able to develop mathematics in new ways and open ourselves to new ideas.

The sea of human endeavour is just as vast as it ever was, but it does not contain monsters. New approaches to knowledge do not challenge our certainty nor destroy everything that has gone before. The post-modern view of mathematics is not to be feared.

In particular it is not to be feared because it has made a huge contribution to our understanding of learning mathematics. The consequences for mathematics education of acknowledging socio-cultural issues have been considerable, and we have a lot for which to thank Bishop, and others, who first brought such issues to our attention.

A more socially constructed picture of mathematics enables us to help new generations of students better understand both the intricacies, and the overall nature, of mathematics.

References

Adam, S. (2004). *Ethnomathematics in the Maldivian curriculum: Trialling an implementation.* Unpublished Phd thesis, The University of Auckland.

Alangui, W. (in process). *Stone walls and water flows: Interrogating cultural practice and its implications to the theory of ethnomathematics.* Unpublished PhD thesis. The University of Auckland.

Alberti, M. (2007). *Interpretación matemática situade de una práctica artesanal. [A situated mathematics interpretaion of an artisan's practice].* Unpublished PhD thesis. Universitat Autònoma de Barcelona.

Ascher, M. (1991). *Ethnomathematics: A multicultural view of mathematical ideas.* New York: Brooks/Cole Publishing Co.

Ascher, M. (2002). The Kolam tradition. *American Scientist, 90*, 57–63.

Barton, B. (1996) Making sense of ethnomathematics: Ethnomathematics is making sense. In S. Lerman (Ed.), *Educational Studies in Mathematics: Special Issue Vol. 31*, 201–233.

Barton, B. (2008). *The language of mathematics: Telling mathematical tales* (Vol. 46). Mathematics Education Library, Springer: New York.

Barton, B., Domite, M., & Poisard, C. (2006). Cultural connections & mathematical manipulations. *For The Learning of Mathematics, 26*(2), 21–24. http://www.math.auckland.ac.nz/Events/2006/ICEM-3/.

Barwell, R., Setati, M., & Barton, B. (Eds.), (2007). *Multilingual Issues in Mathematics Education, a Special Issue of Educational Studies in Mathematics, 65*(2).

Bishop, A. J. (1979). Visualising and mathematics in a pre-technological culture. *Educational Studies in Mathematics, 10*, 135–146.

Bishop, A. J. (1985). The social construction of meaning: A significant development for mathematics education. *For the Learning of Mathematics, 5*(1), 24–28.

Bishop, A. J. (1988). *Mathematical enculturation: A cultural perspective on mathematics education.* Dordrecht, The Netherlands: Kluwer Academic Publishers.

Bishop, A. J. (1990). Western mathematics: The secret weapon of cultural imperialism. *Race & Class, 32*(2), 51–65.

Bishop, A. J. (1994). Cultural conflicts in mathematics education: Developing a research agenda. *For the Learning of Mathematics, 14*(2), 15–18.

Bloor, D. (1976). *Knowledge and social imagery.* London: Routledge & Kegan Paul.

Bloor, D. (1994). What can the sociologist of knowledge say about $2 + 2 = 4$? In P. Ernest (Ed.), *Mathematics, Education and Philosophy: An International Perspective* (pp. 21–32). London: The Falmer Press.

Brown, T. (2001). *Mathematics education and language: Interpreting hermeneutics and post-structuralism.* Dordrecht: Kluwer Academic Publishers.

Contreras L. O., Morales J. F., & Ramírez, J. F. (Eds.). (1999). *Ethnomathematics and mathematics education: Building an equitable future.* Proceedings of First International Conference on Ethnomathematics (ICEM1), CD Rom, Granada, Spain: Universidad de Granada.

Cooke, M. (1990). *Seeing Yolngu, Seeing mathematics*. Northern Territory, Australia: Batchelor College.

Crowe, M. J. (1975). Ten "laws" concerning patterns of change in the history of mathematics. *Historia Mathematica, 2*, 161–166 and 469–470.

Crowe, M. J. (1992). Afterword (1992): A revolution in the historiography of mathematics? In D. Gillies (Ed.), *Revolutions in mathematics*. Oxford: Clarendon Press.

D'Ambrosio, U. (1980). Mathematics and society: Some historical considerations and pedagogical implications. *International Journal of Mathematics Education in Science and Technology, 2*(4), 479–488.

D'Ambrosio, U. (1986). Socio-cultural bases for mathematical education. In M. Carss (Ed.), *Proceedings of the fifth international congress of mathematics education (ICME 5)*. Boston: Birkhauser.

D'Ambrosio, U. (1990). The role of mathematics education in building a democratic and just society. *For the Learning of Mathematics, 10*(3) 20–23.

D'Ambrosio, U. (1992). Ethnomathematics: A research program on the history and philosophy of mathematics with pedagogical implications. *Notices of the American Mathematical Society, 39*(10) 1183–1184.

D'Ambrosio, U. (2001). Mathematics and peace: A reflection on the basis of Western civilization. *Leonardo, 34*(4), 327–332.

Daubin, J. W. (1984). Conceptual revolutions and the history of mathematics. In E. Mendelsohn (Ed.), *Transformation and tradition in the sciences* (pp. 81–103). Cambridge: Cambridge University Press.

Daubin, J. W. (1992). Revolutions revisited. In D. Gillies (Ed.), *Revolutions in mathematics*. Oxford: Clarendon Press.

Davis, B. (1996). *Teaching mathematics: Toward a sound alternative*. Taylor & Francis.

Deregowski, J. B. (1972). Pictorial perception and culture. *Scientific American, 227*, 82–88.

Deregowski, J. B. (1974). Teaching African children depth perception: In search of a method. *Perception 3*, 309–312.

Deregowski, J. B. (1984). *Distortion in art: The eye and the mind*. London: Routledge and Kegan Paul.

Dowling, P. (1998). *The sociology of mathematics education: Mathematical myths/pedagogic texts*. London: The Falmer Press.

Ellerton, N. & Clements, M. (1994). *The national curriculum debacle*. Perth: Meridian.

Frankenstein, M. (1983). Critical mathematics education: An application of Paulo Freire's epistemology. *Journal of Education, 165*, 315–339.

Frankenstein, M. (1998). Goals for a critical mathematical literacy curriculum. In E. Lee, D. Menkart, & M. Okazawa-Rey, *Beyond heroes and holidays: A practical guide to K-12 anti-racist, multicultural education and staff development*, (pp. 306–313). Washington, DC: Network of Educators on the Americas.

Galligan, L. (2001). Possible effects of English-Chinese language differences on the processing of mathematical text: A review. *Mathematics Education Research Journal, 13*(2), 112–132.

Gerdes, P. (1986). How to recognise hidden geometrical thinking? A contribution to the development of anthropological mathematics. *For the Learning of Mathematics, 6*(2) 10–12, 17.

Gerdes, P. (2007). *Adventures in the world of matrices*. Hauppauge NY: Nova Science Publishers.

Halliday, M. A. K. (1975). Aspects of sociolinguistic research, UNESCO, Paris, 1974. A survey paper prepared for *Symposium on the Interactions Between Linguistics and Mathematical Education*, Nairobi, Sept. 1–11, 1974. UNESCO document ED-74/CONF.808/COL.7. Paris: UNESCO.

Hersh, R., & Umland, K. (2005). Mathematical discourse: The link from pre-mathematical to fully mathematical thinking. *Philosophy of Mathematics Ed. Journal, 19*. Online at http://www.people.ex.ac.uk/PErnest/pome19/index.htm.

Hirsch-Dubin, F. (2006). *Evolution of a dream: The emergence of Mayan ethnomathematics and indigenous ways of knowing at a Mayan autonomous school in Chiapas, Mexico*. Paper presented to the 3rd International Conference on Ethnomathematics: Cultural Connections & Mathematical Manipulations. <http://www.math.auckland.ac.nz/Events/2006/ICEM-3/>.

Horsthemeke, K., & Schäfer, M. (2006). *Does "African" mathematics facilitate access to mathematics? Towards an ongoing critical analysis of ethnomathematics in a South African context*. Paper presented to the 3rd International Conference on Ethnomathematics: Cultural Connections & Mathematical Manipulations. <http://www.math.auckland.ac.nz/Events/2006/ICEM-3/>.

Knijnik, G. (1993). An ethnomathematical approach in mathematical education: A matter of political power. *For the Learning of Mathematics, 13*(2) 23–26.

Knijnik, G. (2008). *"Regimes of Truth". On adult peasant mathematics education: An ethnomathematics study*. Paper presented to the Rome Symposium on the Occasion of the 100th Anniversary of ICMI.

Knijnik, G., Wanderer, F., & de Oliveira, C. (2005). Cultural differences, oral mathematics and calculation in a teacher training course of the Brazilian Landless Movement. *Zentralblatt fur Didaktik der Mathematik, 37*(2), pp. 101–108.

Lakoff, G., & Johnson, M. (1980). *Metaphors we live by*. Chicago: University of Chicago Press.

Lakoff, G. & Johnson, M. (1999). *Philosophy in the flesh: The embodied mind and its challenge to Western thought*. New York: Basic Books.

Latterall, C. M. (2004). *Math wars: A guide for parents and teachers*. Praeger Publishers.

Lipka, J. & Adam, S. (2006). *Mathematics in a cultural context: Salmon fishing investigations into probability*. Paper presented to the 3rd International Conference on Ethnomathematics: Cultural Connections & Mathematical Manipulations. <http://www.math.auckland.ac.nz/Events/2006/ICEM-3/>.

Lipka, J., Hogan, M., Parker-Webster, J., Yanez, E., Adams, B., Clark, S., & Lacy, D. (2005). Math in a cultural context: Two case studies of successful culturally based math project. *Anthropology and Education Quarterly, 36*(4), 367–385.

Lipka, J., Sharp, N., Adams, B., & Sharp, F. (2007). Creating a third space for authentic biculturalism: Examples from math in a cultural context?. *Journal of American Indian Education, 46*(3), pp. 94–105.

Loveless, T. (2001). *The great curriculum debate: How should we teach reading and math*. Washington, DC: The Brookings Institution.

Monteiro, Ma. de (Ed.). (2002). *Proceedings of Second International Conference on Ethnomathematics* (ICEM2), CD Rom, Ouro Preto, Brazil: Lyrium Comunacacão Ltda.

O'Halloran, K. L. (2005). *Mathematical discourse: language, symbolism and visual images*. London: Continuum.

Pinxten, R., van Dooren, I., & Soberon, E. (1987). *Towards a Navajo Indian geometry*. Gent: K.K.I. Books.

Powell, A. B. (2002). Ethnomathematics and the challenges of racism in mathematics education. In P. Valero & O. Skovsmose (Eds.), *Proceedings of the Third International Mathematics Education and Society Conference* (Vol. 1, pp. 15–29). Copenhagen: Centre for Research in Learning Mathematics.

Powell, A. B., & Frankenstein, M. (2006). Respecting intellectual diversity: An ethnomathematical perspective. In F. A. Rosamond & L. Copes (Eds.), *Educational transformation: A tribute to Steve Brown* (pp. 161–190). Bloomington, IN: AuthorHouse.

Ray, W. (2001). *The logic of culture: Authority and identity in the modern era*. Oxford: Blackwell Publishers Ltd.

Restivo, S. (1983). *The social relations of physics, mysticism, and mathematics*. Dordrecht: D. Reidel.

Restivo, S. (1992). *Mathematics in society and history: Sociological enquiries*. Dordrecht: Kluwer Academic Publishers.

Restivo, S. (1993). The social life of mathematics. In S. Restivo, J. P. Van Bendegem & R. Fischer (Eds.), *Math worlds: Philosophical and social studies of mathematics and mathematics education*, New York: State University of New York Press.

Rowlands, S. & Carson, R. (2002). Where would formal, academic mathematics stand in a curriculum informed by ethnomathematics? A critical review. *Educational Studies in Mathematics, 50*, 79–102.

Senft, G. (1997). *Referring to space: Studies in Austronesian and Papuan languages*. Oxford: Clarendon Press.

Setati, M. & Adler, J. (2001). Between languages and discourses: Language practices in primary multilingual mathematics classrooms in South Africa. *Educational Studies in Mathematics, 43*(3), 243–269.

Skovsmose, O. (2002). Students' foreground and the politics of learning obstacles. In Monteiro, Ma. de (Ed.), *Proceedings of Second International Conference on Ethnomathematics* (ICEM2), CD Rom, Ouro Preto, Brazil: Lyrium Comunacacão Ltda.

Spengler, O. (1926). *The decline of the west: Form and actuality* (C. F. Atkinson, Trans.). London.

Spengler, O. (1956). Meaning of numbers. In *The world of mathematics* (Vol. IV, pp. 2315–2347). London: George Allen & Unwin Ltd.

Tall, D. (2006). A theory of mathematical growth through embodiment, symbolism and proof. *Annales de Didactique et de Sciences Cognitives*, IREM de Strasbourg. *11*, 195–215.

UNESCO (1975). *Symposium on the interactions between linguistics and mathematical education*, Nairobi, Sept. 1–11, 1974. UNESCO document ED-74/CONF.808/COL.11. Paris: UNESCO.

Vithal, R. & Skovsmose, O. (1997). The end of innocence: A critique of "ethnomathematics". *Educational Studies in Mathematics, 34*, 131–157.

Wilder, R. L. (1950). The cultural basis of mathematics. In *Proceedings of the International Congress of Mathematicians* (Vol 1, pp. 258–271). Cambridge, MA.

Wilder, R. L. (1968). *Evolution of mathematical concepts: An elementary study*. New York: John Wiley & Sons.

Wilder, R. L. (1981). *Mathematics as a cultural system*. Oxford: Pergamon.

Zaslavsky, C. (1991a). Multicultural mathematics education for the middle grades. *Arithmetic Teacher, 38*(6), 8–13.

Zaslavsky, C. (1991b). World cultures in the mathematics class. *For The Learning of Mathematics, 11*(2), 32–36.

Chapter 10
Chinese Culture, Islamic Culture, and Mathematics Education

Frederick Leung

Introduction

Bishop's 1979 paper opened up a fresh perspective on mathematical visualization and cognition in the Papua New Guinea (PNG) culture, a perspective so different from the one that we are used to in the (English) mathematics education literature. The paper illustrated how mathematics learning in a pre-technological culture could be so vastly different from that in a technological society. But looking back 30 years later, is Bishop's paper already out-dated? It may be argued that every culture has gone through these phases of moving from a pre-technological stage to a technological one. The PNG pre-technological culture in the 1970s provided an interesting case because it co-existed with some advanced technological cultures elsewhere, thus allowing even contemporary researchers from those more "advanced" cultures to study it and to expose this transition to the Western world. But the PNG culture might have moved on with more exposure to and interaction with the "Western" culture, and as the majority of the regions around the world have by now finished this transition to a technological society, would Bishop's work represent a mere documentation of the transition at a certain point in our history, and thus be of historical significance only?

As will be argued in the rest of this chapter, Bishop's 1979 paper and his subsequent work have contributed much more to the mathematics education community than just providing an interesting documentation of a certain transition in the history of mathematics education. Bishop did not approach the PNG study with a deficiency model, where the subjects being studied were measured according to the most advanced (Western) education theory, exposing how far behind the subjects being studied were from the ideal. Rather, using the PNG data and data from other studies, Bishop's perceptive observation and critical self-reflection expose and challenge our assumptions on the nature of mathematics, on mathematics cognition, and on mathematics teaching and learning, assumptions which we tend to take for granted since we are all products of our own cultures. This is not easy to achieve,

F. Leung
Chair Faculty of Education, The University of Hong Kong, Pokfulam Road, Hong Kong
e-mail: hraslks@hkucc.hku.hk

P. Clarkson, N. Presmeg (eds.), *Critical Issues in Mathematics Education*,

as "the fish is the last to know water". Bishop was the pioneer "fish", to stay with the metaphor, who became aware of the "water" around us, and he inspired other researchers to study the "water" in addition to studying the aquatic objects. Bishop's work alerts us to the fact that mathematics and its teaching and learning are as much conditioned by culture as any other discipline, and sensitizes us not only to cultural differences that exist, but also to appreciate the strengths rather than the apparent inadequacies in cultures with which we are not familiar.

In 1988 when Bishop was editor of *Educational Studies in Mathematics*, he devoted a special issue of the journal to "Mathematics Education and Culture". Bishop himself contributed a paper (Bishop, 1988), in which he conceptualized mathematics as a sociocultural phenomenon. He discussed the values associated with "Western" mathematics, which he argued was just "one mathematics" among many. Bishop was also one of the major figures behind the organization of the one-day programme (known as the Fifth Day Special Programme) on "Mathematics, Education, and Society" (MES) at the 6th International Congress on Mathematical Education (ICME-6) held in Budapest in the same year. The MES programme was organized around four sub-themes: Mathematics Education and Culture; Society and Institutionalized Mathematics Education; Educational Institutions and the Individual Learner; and Mathematics Education in the Global Village (Keitel, 1989). A large number of cultural factors relevant to mathematics education were examined in the MES papers, and the event prompted further research in this area.

Bishop's 1979 paper said that it aimed at encouraging readers "to consider what might be the implications (of Bishop's work) for the learning and teaching of mathematics" in the readers' "own cultures and countries"'. Responding to Bishop's aim, some mathematics educators began to study the relationship between culture and mathematics education in different countries in the 1980s (Keitel, 1989), and the author of this chapter started to study the interplay between the Confucian culture and mathematics teaching and learning. Like many other students in the East Asian region, the author grew up in a community richly influenced by the Confucian culture and yet received a typical Western mathematics education at school. Many students in this situation had not realized that the Confucian cultural values that underlie the community were relevant to mathematics teaching and learning at all. But if the PNG culture has exerted such a significant influence on PNG students in their learning of mathematics, how can we not expect the Confucian culture to be influencing students in the East Asian region in their mathematics learning as well?

In this chapter, the influence of two major world cultures, namely, the Chinese culture and the Islamic culture, on mathematics teaching and learning will be discussed in order to address Bishop's challenge. The "canonical curriculum" (Howson & Wilson, 1986) studied by students in all parts of the world (including the Chinese world, the Muslim world and the "Western" world) "was developed in Western Europe in the aftermath of the Industrial Revolution" (p. 19) and has a distinctive Greek origin. It is technique-oriented (Bishop, 1991), with a strong emphasis on rationalism. How do the religious Islamic culture and the secular Chinese culture accommodate such a "rational" education?

The Confucian Heritage Culture

Although there are many ethnic groups in China and their cultures differ, Confucianism was the official philosophy and predominant culture throughout most of the Chinese history. The Confucian culture, sometimes referred to as the Confucian Heritage Culture or CHC (Ho, 1991; Biggs, 1996), is still the dominant culture in the country, as well as among most East Asian countries such as Japan, Korea and Singapore. Confucian heritage refers to the legacy due to the famous Chinese scholar Confucius (551–479 BC), perhaps the figure with the greatest influence on the culture of China and its neighbouring countries.

Confucianism is also one of the major cultures in the world. In addition to the one fifth of the world's population in China under its direct influence, East Asian countries and the "overseas" Chinese communities all over the world are broadly under the influence of the Confucian culture. Despite the diverse social and political settings in the CHC communities around the world, it seems that the cultural values they share are relatively stable, so much so that Sun (1983, p. 10) describes the Confucian culture as a "super-stable structure". One possible reason for this relatively stable culture is the Confucian stress on the relationship among family and clan members. For the overseas Chinese for example, it is a common phenomenon that they tend to stay close to each other forming "China towns" with the family units staying intact and upholding a strong relationship among their members. Cultural values are passed on from generation to generation through the family and the Chinese community, so that even when the social and political system changes, these traditional values are still preserved.

Notwithstanding this stable and persuasive culture among a large world population, the influence of CHC on mathematics education has not been studied until recently. Recent interest on this topic among educators was triggered by the superior performance of students from countries which share this culture in international studies of mathematics achievement (e.g., the International Assessment of Educational Progress (IAEP) study (Lapointe, Mead, & Askwe, 1992), studies by Stevenson and Stigler (Stigler, 1992; Stevenson, Lummis, Lee, & Stigler, 1990; Stevenson, Chen, & Lee, 1993), and the Second International Mathematics Study (SIMS) (Robitaille & Garden, 1989)). Much discussion on the results of such international studies, however, did not go beyond comparison of student achievement, but Bishop's work reminds us to look beyond the rankings of countries to the underlying culture that may have an impact on mathematics teaching and learning.

Mathematics Education in CHC Countries

The contribution of mathematicians from CHC, especially from ancient China, has been clearly acknowledged in the field of mathematics (see for example Joseph, 1991). Despite this significant contribution, educators from CHC countries did not seem to have contributed much to the relatively new discipline of mathematics education. In this section, we review different aspects of mathematics education in CHC countries.

Student Achievement

As pointed out in the last section, recent interest in CHC countries stems from the superior performance of students from these countries in international comparative studies of mathematics achievement. For example, East Asian students have consistently out-performed their counterparts around the world in the Trends in International Mathematics and Science Study or TIMSS[1] (Beaton et al., 1996; Mullis et al., 1997, 2000; Mullis, Martin, Gonzales, & Chrostowski, 2004) and the *Organization for Economic Cooperation and Development* (OECD) Programme for International Student Assessment study or PISA (OECD, 2001, 2003, 2004). In TIMSS 1995, the four countries[2] or systems that topped the list in mathematics achievement in middle and primary school years were Singapore, Korea, Japan and Hong Kong (Beaton et al., 1996; Mullis et al., 1997). These four were also the only countries that could be classified as falling under the influence of CHC out of the more than 40 countries that participated in the study. Similarly, in TIMSS 1999, there were only five CHC countries (the four in TIMSS 1995 plus Chinese Taipei), but they topped the 38 countries in grade eight mathematics (Mullis et al., 2000). Furthermore, the difference in the level of achievement between these high performing CHC countries and many of the other TIMSS countries was rather substantial. In TIMSS 1999 for example, all the five CHC countries were more than three standard deviations above the lowest scoring country, and more than one standard deviation above 15 of the countries (Mullis et al., 2000). Similar results were obtained in the TIMSS 2003 study (Mullis et al., 2004) as well as the PISA studies (OECD, 2001, 2003, 2004). These results are also consistent with those of the earlier studies mentioned above (IAEP, Stevenson and Stigler's studies and SIMS), and so it seems that the superior performance of East Asian students is rather stable over time.

Actually, when the TIMSS results were first released in 1996, the superior performance of students from the East Asian countries (except possibly for Japan) came as a surprise to many mathematics educators. They had expected that countries such as Russia, Hungary and US would outperform the other countries. For the East Asian countries, the literature had indicated that their teaching was very traditional and out-dated. Lessons were teacher dominated, content oriented and examination driven. Student involvement was minimal, and memorization and rote learning was

---

[1] TIMSS was the abbreviation of the *Third International Mathematics and Science Study* when the study was conducted in 1995 and 1999. From 2000 onwards, the study has been renamed as *Trends in International Mathematics and Science Study*.

[2] Some participants in TIMSS are education systems which are not countries (e.g. Hong Kong, Flemish speaking Belgium), but for the ease of presentation in this chapter, instead of saying "countries/systems" every time, the generic term "countries" will be used to refer to countries or systems.

encouraged (Brimer & Griffin, 1985; Biggs, 1994; Leung, 1995). How could these countries top the list of TIMSS and PISA countries in mathematics achievement?

Language as a Cultural Explanation of Cognition and Achievement

In the ICME-6 MES programmes mentioned at the beginning of the chapter, at least five of the papers in MES touched on the influence of language in mathematics learning (Kurth; Leung; Lorcher; Watson; and Zepp; see Keitel, 1989), but only one of them related to a CHC language (Chinese). Another study touching upon the relationship between the Chinese language and mathematics learning was a study conducted by Lin (1988) in Taiwan. Lin's study was a replication of the CSMS study in the UK (Hart, 1981), which looked into, among other issues, the influence of language on children's understanding of algebraic symbols. One of the surprising findings of Lin's study was that "the most difficult item for the English students proved to be the easiest item ... for Taiwan students" (p. 476). Lin explained the results in terms of the different features between the English and Chinese languages.

In a more recent study, Leung and Park (2007) found that language had a direct impact on students' recognition of simple geometric figures and their understanding of the concepts involved. The way names are coined to designate elementary geometric figures differs in the languages of English, Chinese and Korean, and it was found that the language use exerted an influence on students' understanding of the figures. The same problem may arise for other areas of mathematics.

Instructional Practice

While the influence of language on mathematics learning is an interesting and important factor to consider in understanding mathematics cognition and in explanation of student achievement, language is more or less a "given" in the teaching and learning process, and as such is a factor that is hard to change. In considering the interplay between culture and mathematics teaching and learning, more attention has been paid to instructional practices in the classroom as a possible explanation of student achievement.

The international studies such as TIMSS and PISA mentioned above included in their instruments, questionnaires that studied the "implemented curriculum" of the countries concerned. Actually, the TIMSS organizers stressed that the main purpose of these international studies was not to compose a league table of student performance in different countries (although this is the part that the media is most interested in), but to identify factors that explain student achievement, and instructional practice is clearly one such important factor. Results of these international studies, however, failed to identify any instructional practices worldwide that are related with high student achievement. While this may be disappointing to some educators, this result should not be surprising at all when considered from a socio-cultural perspective, a perspective which is in line with the approach promulgated by Bishop. The same reported instructional activity (in response to a questionnaire) may mean completely different things in different cultures. How can we expect a questionnaire to be able to capture the complexities of classroom teaching?

Of course, there are other limitations to the use of questionnaires to study instructional practices in a cross-cultural setting. The International Association for the Evaluation of Educational Achievement (IEA), the organizer of TIMSS, was also aware of these limitations, and so alongside TIMSS 1995, a video study that examined instructional practices in grade 8 mathematics was conducted for the three countries of Germany, Japan and the United States. Altogether 231 lessons were videotaped and analyzed, and results of the study (Stigler & Hiebert, 1999) attracted much attention from the mathematics education community and the public at large. A striking finding of the Study was that the instructional practices in the classrooms of Japan, the only CHC country in the study, differed drastically from those in the two Western countries of Germany and the United States. Was this difference a result of the idiosyncrasies of Japan, or was this an expression of the CHC culture?

Because of the success of the TIMSS 1995 Video Study, the study was conducted again in 1999, and this time seven countries or systems (Australia, Czech Republic, Hong Kong, Japan,[3] the Netherlands, Switzerland and the US) participated in the study (Hiebert et al., 2003). Since Japan did not collect any data in 1999 and the only other CHC system was Hong Kong, it is hard to generalize from the results of the Study whether there is a CHC pedagogy in mathematics. The results however still indicated a certain degree of differences in instructional practice between Hong Kong and Japan on one hand, and the "Western" countries on the other (with the possible exception of the Czech Republic, which in a number of measures performed quite similarly to the two CHC countries). One salient feature of the CHC classroom was the dominance of the teacher in the teaching and learning process, but the findings of the study also showed that high quality teaching and learning could still take place even in a teacher directed classroom.

This finding echoes well with the results of a video study of mathematics classrooms in Hong Kong and Shanghai by Huang and Leung (2004), and those of another large scale multi-national video study, the Learner's Perspective Study (LPS). Huang and Leung (2004) also reported that the CHC classrooms in their study were characterized by teacher dominance. Yet there was active student engagement and much emphasis on exploration of mathematics through practicing exercises with variation in these classrooms. In analyzing the LPS data for Korea, Park, and Leung (2006) found that "active student engagement is still possible in a classroom where the class size is large and the activities are dominated by the teacher" (p. 257).

In an earlier paper, Leung (1999) discussed the traditional Chinese views of mathematics and education which might have an impact on the classroom practices

[3] Japan did not collect video data for mathematics in 1999, but the Japanese data for the TIMSS 1995 video study were re-analyzed using the 1999 methodology in some of the analyses.

in the contemporary Chinese classroom. In another paper, Leung (2001) extended the argument from the Chinese classroom to the East Asian classroom and identified features of East Asian mathematics education in contrast to features in the West. Leung (2005) argued that these East Asian classroom practices are "deeply rooted in the underlying cultural values of the classroom and the wider society" (p. 212).

Another interesting finding evolved from the TIMSS 1999 Video Study results for Hong Kong and Japan. In examining the quantitative analysis and the qualitative analysis of the same sets of data, it was found that two rather different pictures of the CHC classroom were portrayed by the two analyses (Leung, 2005). This discrepancy may pose some challenges to the validity of video studies in general, but a closer look at the two methodologies reveals that perhaps different aspects of the reality were portrayed by the two analyses. This has methodological implications for the analysis of video data, but more pertinent to the discussion here is that the seemingly contradictory findings of the two analyses point to the complexity of classroom events. As Bishop (1991) pointed out, the classroom is just part of the larger cultural community, so superficial quantitative measures of classroom events may not reveal the subtlety of the activities that took place under the given cultural context. In other words, the same superficial activities (e.g., the number of words spoken by the teacher, the number and kind of questions the teacher asks in the lesson, the number of problems solved, etc.) may mean very different things in different cultures. To draw meaningful conclusions from the video data, expert judgment is needed, but of course that may render the subjective data analysis unreliable.

Teacher Knowledge

From classroom practices, one naturally turns to the teacher himself or herself for explanations of both student achievement and instructional practices. To what extent are instructional practices, and in turn student achievement, attributable to teacher's knowledge in mathematics and pedagogy, and if teachers' knowledge and pedagogy differ, can these be explained by the different teacher education programmes they went through? More importantly, if teacher knowledge and teacher education do differ from country to country, how much of these differences can be attributed to the different cultures of the countries concerned?

These are important questions triggered by this attention to culture, and they no doubt will be major issues of concern in mathematics educational research in the years to come (IEA for example has just launched a Teacher Education and Development (Mathematics) project, for which teacher education programmes and teacher competence are to be compared on an international scale). Some initial results on teacher knowledge in CHC countries however have already started to emerge.

Studies by Ma (1999) and An, Kulm, and Wu (2004) both pointed to the obvious conclusion that teachers' knowledge made a major impact on their teaching practice. For example, although the primary purpose of Ma's study was not a comparison of teacher knowledge, her study showed that the Shanghai elementary mathematics teachers in her sample had a "profound understanding of fundamental mathematics", and as a consequence their instructional practices differed substantially from the American teachers who did not have such profound knowledge (Ma, 1999). In a replication of Ma's study in Hong Kong and Korea (Leung & Park, 2002), it was found that most of the Hong Kong and Korean teachers in the study had a good grasp of the underlying concepts of elementary mathematics. The teachers were also found to be proficient in mathematics calculations, although compared to the Shanghai teachers in Ma's study, they were weak in their ability to guide students in genuine mathematical investigations.

An et al.'s comparative study on middle school mathematics teachers in China and the U.S. also found that the pedagogical content knowledge of mathematics teachers in the two countries differed remarkably. Consequently, in their teaching, the Chinese teachers emphasized developing procedural and conceptual knowledge through reliance on traditional and more rigid practices, while American teachers gave emphasis to a variety of activities designed to promote creativity and inquiry in attempting to develop students' understanding of mathematical concepts (An et al., 2004).

Summary

We can see that the study by Bishop in PNG and Bishop's subsequent stress on the socio-cultural aspect of mathematics education have triggered a sensitivity among educators to interpret student achievement, instructional practice and teacher education in a major world culture from a cultural perspective. Of course culture is not the only explanatory variable for these educational phenomena, but the discussion above shows that it is definitely an important variable, and one that was pretty much ignored prior to Bishop's effort in this area of study.

The Islamic Culture

Bishop might not have CHC in mind when he reported on the PNG study in 1979 (in fact the term CHC was coined only in the early 1990s), nor even when he wrote the book *Mathematical Enculturation* in 1991. But he did have another major culture in mind (although this was only marginally mentioned in *Mathematical Enculturation* (see Footnote 1 of Chapter 4)), that of the Islamic culture. Back in the 1980s, Bishop in a personal communication to the author, mentioned that he was working with a group of Islamic educators who were visiting Cambridge (where Bishop was working at that time). Bishop also gave a talk to Islamic scholars at a conference in Iran in 1997, where he commented on the contribution of the Islamic culture to mathematics and mathematics education.

To take up Bishop's interest in this area, what follows is a brief review of the work that has been undertaken on the relationship between the Islamic culture and mathematics education. Of course, as in the case of CHC, the review is not meant to imply that such work is a direct result of Bishop's advocacy of attending to cultural influences of mathematics education. The survey represents more a tribute to Bishop's work in this area by the present author.

Mathematics Education in the Islamic Culture

Islamic Contribution to Mathematics

Science and mathematics in the modern world have benefited much from the contributions of the Islamic culture, although sometimes these contributions are not much noticed (Taher, 1997). After the September 11, 2001, event in New York, there has been some renewed interest in the Western world in the Islamic culture, including interest in the education under this culture (Griffin, 2006). However, work specifically on Islamic views on, or contribution to, mathematics education is relatively rare.

In contrast, there have been many studies on Islamic contributions to mathematics, particularly computational mathematics. Scholars on Islamic mathematics often point to the "golden age of the Islamic era", roughly between the 10th and 16th century AD in the Arabic world. Works of famous Islamic mathematicians such as Al-Khwarizmi (born about AD 790), Al-Biruni (born about AD 973), Umar al-Khayyami (born about AD 1048) and Al-Kashi (born late 14th century) (see Berggren, 1986) are rather well-known to many mathematicians.

Mathematics Education

As far as mathematics education is concerned, Abdeljaouad (2006) completed a very thorough study on the history of mathematics teaching in the Islamic culture. He looked into issues such as the status of mathematics in the Islamic culture, the kind of mathematics taught, the teachers of mathematics and the institutions where mathematics was taught, the textbooks used, the pedagogy employed, etc. Abdeljaouad pointed out at the outset that Islamic education is religious in nature. "Islamic education has always been primarily religious education, in the sense that it is explicitly intended to preserve the religious tradition from which Islamic communal life springs" (De Young, 1986, p. 3, cited in Abdeljaouad, 2006, p. 634). At the same time, there is a strong urge to seek knowledge in the culture, as evidenced by such Islamic mottos as "seek knowledge even if it is in China", "asking and demanding for knowledge is an obligation for every man and woman" and "seeking knowledge from cradle to grave is an obligation for every human being" (Gooya, 2008, p. 1). It is well known that Islamic scholars throughout their history have been very active in learning from foreign sources. During the "golden age of the Islamic era", Islamic scholars translated many texts from ancient Greece, India, and old Persia, and Islamic mathematics and mathematics education were evidently much influenced by works from these cultures. Islamic works in this era in turn formed the "scientific pillars of the Renaissance in the Western world" (Gooya, 2008, p. 1).

Berggren (1990) studied how proof was conceived in medieval Islamic mathematics. He found that "mathematics flourished in the forms of methods and techniques rather than that of theorems and proofs" (p. 47). While some medieval Islamic mathematics satisfied the modern criteria for proof (Hanna, 1983), many were simply a "body of techniques" (for example, for astronomical calculations) "for whose validity no argument was ever given" (p. 40). The main purpose for such mathematics was for solving practical problems. In this regard, medieval Islam mathematics was rather similar to ancient Chinese mathematics (see Joseph, 1991), but evidence seems to point to the fact that the Islamic mathematics tradition was more diverse, with practically oriented and curiosity driven mathematicians coexisting (Gooya, 2008). Both "'theoretical' mathematics necessary to understand the world and 'practical' mathematics used to solve the problems of everyday life" were taught and treasured (Abdeljaouad, 2006, p. 634). More importantly, Berggren (1990) pointed out that Islamic mathematics often included an analysis of the theorem or mathematical result to be proved, perhaps to convince "the reader that the proof followed is the most natural one" (p. 43). A possible reason for this inclusion of an analytic discussion of the mathematical results may be for the purpose of training (advanced) students. Muslim mathematicians were known to have "a lively concern for pedagogy" (p. 43). These clearly have implication for mathematics education in the contemporary world.

Islamic Contribution to Pedagogy

Makdisi (1981) noted that "the development of the memory is a constant feature of medieval education in Islam" (p. 99), but memorizing was "not meant to be unreasoning rote learning", it was to be "reinforced with intelligence and understanding" (p. 103). This is rather similar to the theory of variation as espoused by Marton (Marton & Booth, 1997) and Gu (1994) for mathematics teaching and learning in China (see Huang & Leung, 2004).

Rofagha (2006) investigated the mathematical and pedagogical contribution of Sheikh Bahai, a 16th century Middle Eastern Islamic teacher and scholar. Through detailed analysis of Sheikh Bahai's mathematical work, Rofagha presented several of its approaches that he argued to be beneficial to modern elementary and secondary mathematics education.

Golafshani (2005) examined Iranian mathematics teachers' beliefs about the nature of mathematics and its teaching and learning within the two theoretical frameworks of Absolutist/Traditional and Constructivist/Non-Traditional. The results showed that contrary to findings in other developing countries, Iranian secondary teachers expressed greater support for non-traditional than traditional beliefs about mathematics, mathematics teaching and mathematics learning. Golafshani attributed the differences between this study and previous research on teacher beliefs in developing countries to historical, social and cultural features of Iranian educational policy and practice.

Much of the computational mathematics that the modern student is learning is due to the contribution of Islamic mathematicians. For Islamic students, they should rightly be proud of this significant contribution of their ancestors. Another area of mathematics that is much prized by the mathematics teacher in the modern classroom is the beautiful geometric patterns in Islamic art, especially those associated with the Islamic religion. Individual teachers have reported on how they capitalized on this aspect of the Islamic culture in teaching mathematics, especially in countries where students come from diverse backgrounds (see for example Zaslavsky, 1993, 2002). Abas (2001) pointed out that the beautiful Islamic geometric patterns were excellent materials to teach the idea of symmetry at all levels, from properties of simple geometric figures to tessellation, transformation, and to providing "a visual gateway for the teaching of abstract notions of Group Theory at the university level" (Abas, 2001, p. 53).

Conclusion

In the chapter, I have traced how, triggered by Bishop's study of the influence of culture on mathematics education, attention has been directed to the interplay between CHC and mathematics teaching and learning. The chapter also touched on the relationship between the Islamic culture and mathematics education. In contrast to the PNG culture on which Bishop reported in his 1979 paper, clearly these two cultures can by no means be labelled as pre-technological. But a closer and critical look at how mathematics was taught and learned in these cultures has shown that even in these technological societies, mathematics teaching and learning were conceived and practiced rather differently from other technological societies in the West. Since Bishop argued that mathematics education is a kind of "enculturation" (Bishop, 1991), I leave it to the readers to draw their own conclusion on the values into which teachers from these cultures "enculturate" their students, in contrast to the typical Greek values underlying the Western mathematics culture.

References

Abas, S. J. (2001). Islamic geometrical patterns for the teaching of mathematics of symmetry. *Symmetry in Ethnomathematics, 12*(1–2), 53–65.

Abdeljaouad, M. (2006). Issues in the history of mathematics teaching in Arab countries. *Paedagogica Historica, 4* (4 & 5), 629–664.

An, S., Kulm, G., & Wu, Z. (2004). The pedagogical content knowledge of middle school mathematics teachers in China and the U.S. *Journal of Mathematics Teacher Education, 7*, 145–172.

Beaton, A. E., Mullis, I. V. S., Martin, M. O., Gonzalez, E. J., Kelly, D. L., & Smith, T. A. (1996). *Mathematics achievement in the middle school years*. Boston: Centre for the Study of Testing, Evaluation and Educational Policy, Boston College.

Berggren, J. L. (1986). *Episodes in the mathematics of medieval Islam.* New York: Springer-Verlag.

Berggren, J. L. (1990). Proof, pedagogy, and the practice of mathematics in Medieval Islam. *Interchange, 21*(1), 36–48.

Biggs, J. B. (1994). What are effective schools? Lessons from East and West (The Radford memorial lecture). *Australian Educational Researcher, 21*, 19–39.

Biggs, J. B. (1996). Western misconceptions of the Confucian-Heritage Learning Culture. In D. A. Watkins & J. B. Biggs, (Eds.), *The Chinese learner*. Hong Kong: Comparative Education Research Centre.

Bishop, A. J. (1988). Mathematics education in its cultural context. *Educational Studies in Mathematics, 19*(2), 179–191.

Bishop, A. J. (1991). *Mathematical enculturation.* Dordrecht: Kluwer Academic Publishers.

Brimer, A. & Griffin, P. (1985). *Mathematics achievement in Hong Kong secondary schools.* Hong Kong: Centre of Asian Studies, the University of Hong Kong.

De Young, G. (1986). The Khulasat al-Hisab al-Hisab of Baha al-Din al-Amili and the Dar-i-Nizami in India. *Indian Society for History of Mathematics, 8*(1–4).

Golafshani, N. (2005). *Secondary teachers' professed beliefs about mathematics, mathematics teaching and mathematics learning: Iranian perspective*, PhD thesis, University of Toronto.

Gooya, Z. (2008). *Speculations on mathematics education in the golden age of Islamic era in Iran*, personal communication.

Griffin, R. (Ed.). (2006). *Education in the Muslim world: Different perspectives.* Oxford, UK: Symposium Books.

Gu, L. (1994). *Theory of teaching experiment – The methodology and teaching principle of Qingpu.* Beijing: Educational Science Press.

Hanna, G. (1983). *Rigorous proof in mathematics education.* Toronto: OISE Press.

Hart, K. M. (1981). *Children's understanding of mathematics: 11–16.* London: John Murray.

Hiebert, J., Gallimore, R., Garnier, H., Givvin, K. B., Hollingsworth, H., Jacobs, J., Chui, A. M. Y., Wearne, D., Smith, M., Kersting, N., Manaster, A., Tseng, E., Etterbeek, W., Manaster, C., Gonzales, P., & Stigler, J. (2003). *Teaching mathematics in seven countries. Results from the TIMSS 1999 video study.* Washington DC: National Center for Education Statistics.

Ho, D. Y. F. (1991). *Cognitive socialization in Confucian heritage cultures.* Paper presented at the Workshop on Continuities and Discontinuities in the Cognitive Socialization of Minority Children, US Department of Health and Human Services. Washington DC, June 29–July 2.

Howson, G. & Wilson, B. (1986). *School mathematics in the 1990s.* Cambridge: Cambridge University Press.

Huang, R. & Leung, F. K. S. (2004). Cracking the paradox of the Chinese learners – Looking into the mathematics classrooms in Hong Kong and Shanghai. In L. Fan, N. Wong, J. Cai, & S. Li (Eds.), *How Chinese learn mathematics: Perspectives from insiders.* New Jersey: World Scientific.

Joseph, G. G. (1991). *The crest of the peacock: Non-European roots of mathematics.* London: I. B. Tauris.

Keitel, C. (Ed.). (1989). *Mathematics, education, and society.* Paris: UNESCO.

Lapointe, A. E., Mead, N. A., & Askwe, J. M. (Eds.), (1992). *The International Assessment of Educational Progress Report No. 22-CAEP-01: Learning mathematics.* New Jersey: The Centre for the Assessment of Educational Progress, Educational Testing Service.

Leung, F. K. S. (1995). The mathematics classroom in Beijing, Hong Kong and London. *Educational Studies in Mathematics, 29*, 297–325.

Leung, F. K. S. (1999). The traditional Chinese views of mathematics and education: Implications for mathematics education in the new millennium. In C. Hoyles, C. Morgan & G. Woodhouse (Eds.), *Rethinking the mathematics curriculum.* London: Falmer Press.

Leung, F. K. S. (2001). In search of an East Asian identity in mathematics education. *Educational Studies in Mathematics, 47*, 35–52.

Leung, F. K. S. (2005). Some characteristics of East Asian mathematics classrooms based on data from the TIMSS 1999 Video Study. *Educational Studies in Mathematics, 60*(2), 199–215.

Leung, F. K. S. & Park, K. (2002). Competent students, competent teachers? *International Journal of Educational Research, 37*, 113–119.
Leung, F. K. S. & Park, K. (2007). The influence of language on students' learning of mathematics. Paper presented at the *Third Conference on Mathematics Curriculum and Teaching Across the Straight*, May, Chongqing.
Lin, F. L. (1988). Social differences and their influences on children's mathematics understanding. *Educational Studies in Mathematics, 19*, 471–497.
Ma, L. (1999). *Knowing and teaching elementary mathematics: Teachers' understanding of fundamental mathematics in China and the United States*. Mahwah, NJ: Erlbaum.
Makdisi, G. (1981). *The rise of colleges, institutions of learning in Islam and the West*. Edinburgh: Edinburgh University Press.
Marton, F. & Booth, S. (1997). *Learning and awareness*. Mahwah, NJ: Lawrence Erlbaum.
Mullis, I. V. S., Martin, M. O., Beaton, A. E., Gonzalez, E. J., Kelly, D. L., & Smith, T. A. (1997). *Mathematics achievement in the primary school years*. Boston: Center for the Study of Testing, Evaluation and Educational Policy, Boston College.
Mullis, I. V. S., Martin, M. O., Gonzalez, E. J., Gregory, K. D., Garden, R. A., O'Connor, K. M., Chrostowski, S. J., & Smith, T. A. (2000). *TIMSS 1999 international mathematics report*. Boston: International Study Center, Lynch School of Education, Boston College.
Mullis, I. V. S., Martin, M. O., Gonzalez, E. J., & Chrostowski, S. J. (2004). *TIMSS 2003 international mathematics report*. Boston: International Study Center, Lynch School of Education, Boston College.
Organisation for Economic Co-operation and Development (2001). *Knowledge and skills for life: First results from PISA 2000*. Paris: OECD Publications.
Organisation for Economic Co-operation and Development (2003). *Literacy skills for the world of tomorrow – Further results from PISA 2000*. Paris: OECD Publications.
Organisation for Economic Co-operation and Development (2004). *Learning for tomorrow's world – First results from PISA 2003*. Paris: OECD Publications.
Park, K. & Leung, F. K. S. (2006). Mathematics lessons in Korea: Teaching with systematic variation. In D. J. Clarke, C. Keitel, & Y. Shimizu (Eds.), *Mathematics classrooms in twelve countries: The insider's perspective* (pp. 247–261). Rotterdam: Sense Publishers.
Robitaille, D. F. & Garden, R. A. (Eds.). (1989). *The IEA study of mathematics II: Contexts and outcomes of school mathematics*. Oxford: Pergamon Press.
Rofagha, N. (2006). *An investigation of the mathematical and pedagogical contribution of Baha al-Din Muhammad ibn Husayn Amili (aka Sheikh Bahai), a sixteenth century Middle Eastern teacher and scholar*. Unpublished PhD thesis, Columbia University.
Stevenson, H. W., Lummis, M., Lee, S. & Stigler, J. (1990). *Making the grade in mathematics: Chinese, Japanese and American children*. Reston, VA: National Council of Teachers of Mathematics.
Stevenson, H. W., Chen, C., & Lee, S. Y. (1993). Mathematics achievement of Chinese, Japanese, and American children: Ten years later. *Science, 25*(9), 53–59.
Stigler, J. W. (1992). *The learning gap*. New York: Touchstone.
Stigler, J. W. & Hiebert, J. (1999). *The Teaching Gap*. New York: Free Press.
Sun, L. K. (1983). *The "Deep Structure" of Chinese culture*. Hong Kong: Chap Yin Co. [in Chinese].
Taher, M. (Ed.). (1997). *Educational developments in muslim world*. New Delhi: Anmol Publications Pvt. Ltd.
Zaslavsky, C. (1993). *Multicultural mathematics*. Portland, ME: J. Weston Watch.
Zaslavsky, C. (2002). Exploring world cultures in math class. *Educational Leadership, 60*(2), 66–69.

Section V
Social and Political Aspects

Bishop had spent a formative 3 years as a graduate student at Harvard University in the mid 1960s, immediately after completing his initial university qualifications in the UK. For much of this time in the USA he taught part time in schools. Then as now, most graduate students are always in need of cash. However being both a graduate student and a teacher, he lived what were not only broadening experiences at that time, but they proved to be useful experiences to draw on much later. By the 1990s Bishop was trying to position what was happening in England in mathematics education, site of his ongoing and immediate experience, to what was happening throughout the world. This process of course had begun much earlier, but he seems to start to draw together his thinking in a rather decisive manner at this time.

The key article for this section takes stock of a number of threads in Bishop's research, a decade or so after his formative visit to Papua New Guinea in 1977. Clearly by the early 1980s he had met many overseas colleagues and developed an extensive international network, somewhat harder to do then before the ubiquitous email conversations we have now, and even before the explosion of cheap air fares. Part of this network was working with the international organisations of the *International Congress of Mathematical Education* (ICME) and the *International Group for the Psychology of Mathematics Education* (PME), editing of the journal *Educational Studies in Mathematics* for some years, being a crucial player in the BACOMET group (see Chapter 2, p. 18 and Chapter 11, p. 185 of this volume), and editing a mathematics education research book series for Kluwer (now Springer) which brought him into contact with many colleagues across the world. It is no wonder that within this environment, for someone who wanted change to occur in classrooms, schools, and indeed the wider educational systems, Bishop needed to develop, and did, a political stance, and recognised what action could flow from that.

In particular the key article is a review of two documents that were published in the USA nearly 10 years after what was probably the crucial document, An *Agenda for Action*. The "Agenda" led the break with the so called 1960s "new maths" to what we regard now as the "normal" mathematics education agenda. Bishop was invited by the editors of the Harvard Educational Review as a knowledgeable "outsider" to comment on the "new way" that began with the "Agenda", but was now reaching something of a watershed with a political document and a curriculum document, both published at about the same time by overlapping groups of scholars.

Bishop's review is interesting for at least two "down to earth" approaches that he takes. He asserts that mathematics is not everything, nor even necessarily the core, of a child's education. Mathematics, like every other component of education, must justify why it should be taken seriously as part of that education. Secondly he is quite clear that there does need to be political action for change to occur. It is no longer possible for academics to simply state a case and expect the policy makers to unquestionably accept their advice. Although articulating the position for change is an essential element, Bishop recognised the need for the long hard work of convincing the power brokers of the need for change, and what form that change should take. Although he was writing of the situation in the USA, he continued to articulate this position throughout his work.

In reflecting on this key article, Keitel and Vithal in a joint chapter start with Bishop's contributions to this critical issue for mathematics education, canvassing many more of his published works than just the key article of this section, but going on to examine the issue from a number of perspectives. Contrasting their own countries' perspectives, delving back through history, deconstructing what was to begin with an important rallying call of "Mathematics for all" but has since been shown to be quite hollow, and finishing back in the classroom with teachers, the authors in a few pages paint a picture that is quite pessimistic but at the same time challenging to our community.

An Additional Bishop References Pertinent to This Issue

Bishop, A.J. (1994). Cultural conflicts in mathematics education: Developing a research agenda. *For the Learning of Mathematics, 14*(2), 15–18.

Chapter 11
Mathematical Power to the People

Alan J. Bishop, University of Cambridge

EVERYBODY COUNTS: A REPORT TO THE NATION ON THE FUTURE OF MATHEMATICS EDUCATION
by the Mathematical Sciences Education Board, the Board on Mathematical Sciences, and the Committee on the Mathematical Sciences in the Year 2000. *Washington, DC: National Academy Press, 1989. 144 pp. $7.95.*

CURRICULUM AND EVALUATION STANDARDS FOR SCHOOL MATHEMATICS
prepared by the Working Groups of the Commission on Standards for School Mathematics of the National Council of Teachers of Mathematics. *Reston, VA: National Council of Teachers of Mathematics, 1989. 258 pp. $25.00.*

It is occasionally interesting and important to find out what the neighbors are doing–and in education it is no different. It is particularly interesting and important to do this at a time when educational systems around the world are trying to respond to a variety of common problems–identified and analyzed by Philip Coombs as stemming from changes in the economic, political, and demographic environments in the 1970s and 1980s.[1] Therefore, it has been extremely instructive to cast a critical eye over two recent reports initiated by the mathematics education community in the United States. In the United Kingdom, we have also been engaged in a reform process throughout the 1980s that has resulted in a national agenda, albeit of a rather different nature. Our social contexts and educational systems may be very different, but the political pressures and the professional concerns are similar. My sense is that both the U.K. and U.S. reform movements have learned and incorporated practices from elsewhere. This essay should, therefore, be viewed as an attempt to increase our mutual understanding about the reform process in mathematics education.

The two reports being reviewed here are the products of two distinct communities that do not always see eye-to-eye in any country–mathematical science and mathematics education. In this case, however, the correlation and the collaboration have been remarkable. *Everybody Counts*, produced by the National Research Council

---

[1] Philip H. Coombs, *The World Crisis in Education–The View from the Eighties* (New York: Oxford University Press, 1985).

Harvard Educational Review Vol. 60 No. 3 August 1990

0017-8055/90/0800-0357$01.25/0

(NRC) through its Mathematics Sciences Education Board (MSEB), sets the scene, so to speak, for the *Curriculum and Evaluation Standards for School Mathematics* (generally referred to as *Standards*) document produced by the National Council of Teachers of Mathematics (NCTM). The *Standards* report augments and amplifies with exemplary curricular material the more general and, to some extent, idealized structure argued for in *Everybody Counts*. In my view, the reports are very much a pair. Yet the number of people who will read them as a pair is likely to be small, and will probably be limited to those working within the mathematics education community. This is unfortunate, since they need a wider audience in order to fully realize their goals.

The stated reason for the collaboration is clear: both communities share a concern over the state of the nation's mathematical education at all levels and sense an opportunity to rewrite the reform agenda. The NCTM's *An Agenda for Action* began the process, but it was *A Nation at Risk* and the follow-up reports that framed an educational and political climate within which it was possible to delegitimize the minimal competency ideology of minimal expectations and minimal demands.[2] The time appears to be ripe for the mathematics education community to present a different vision of mathematics teaching, an "excellence" vision of which professional mathematicians and the rest of the nation can be proud. This vision is not elitist – far from it – the goal is nothing less than an excellent mathematical education for all. An impossible goal, perhaps? Certainly a high ideal, but I feel that it is undoubtedly worth striving for – unlike that of the minimal competency and back-to-basics movements, which settled for the lowest common denominator of achievement as their goal. Minimalist arguments may be easy to defend, but in my experience they rarely inspire students or teachers.

Everybody Counts lays out the manifesto. The National Research Council and the Mathematics Sciences Education Board marshal an array of research findings, argument, and opinion concerning student achievement and educational practice in mathematics at all levels, and recommend a comprehensive program of development to tackle the "crisis."[3] A tenuous link is made in *Everybody Counts* between the decline in general student achievement in mathematics and the decline in the

---

[2] National Council of Teachers of Mathematics, *An Agenda for Action: Recommendations for School Mathematics of the 1980's* (Reston, VA: NCTM, 1980); National Commission on Excellence in Education, *A Nation at Risk: The Imperative for Educational Reform* (Washington, DC: GPO, 1983). See also A. Harry Passow, "Reforming Schools in the 1980's: A Critical Review of the National Reports" (New York: ERIC Clearinghouse on Urban Education, Teachers College, Columbia University, 1984), and a critical review of some of these reports by George M. A. Stanic in the *Journal for Research in Mathematics Education, 15* (1984), 383-389.

[3] This reviewer must confess to having a problem coping with the extreme language of reports like this. We in the United Kingdom do have our extreme writers, of course, but it does appear to be part of the American way that, for whatever reason, one has to use words and phrases like "crisis," "disaster," "future in jeopardy," "nation at risk," and so on, if one wants to make any impact on the American psyche. The criticisms, both overt and applied, that accompany such rhetoric – that firstly it is the object of concern which is *solely* responsible for the "disaster," and secondly that it is the *whole* "object" that is responsible–are inherently false, and, to my mind, the arguments

competitiveness of the U.S. economy. This is an extraordinarily difficult line to argue. The evidence, real or imagined, is at best correlational, and the arguments rest more on basic and usually unexamined political tenets than on an understanding of educational realities. The corollary, that putting mathematics education to rights will *consequently* improve the U.S. economy, is an argument designed more for its appeal to the politician's and the business chiefs memo-writer than for its ability to convince the rest of the education profession. One must, therefore, read this report as a political document, because it is concerned with policy, with persuasion, and with power. It is written by an influential body and is intended to be read by other influential bodies: policymakers, industrial leaders, politicians, and people with local educational influence. It *had* to be written starkly; its subtitle is *A Report to the Nation on the Future of Mathematics Education.*

From this perspective I find the document impressive. It is attractively presented and liberally sprinkled with photographs and diagrams. Its simple language, with highlighted quotations, section headlines, and uncluttered prose (notes are at the back), conveys the message in a direct and no-nonsense manner. This is no academic tome that could be dismissed by the business and political communities as over-intellectualized. The arguments used are those which politicians in both our countries seem to value, where the problem is clear, the facts are undisputed, the simple statistical graphs catch the essential trends, and the logic is "cause and effect." The mood is macho and up-beat. The problem *is* solvable and "we" know how to solve it.[4]

A sample of highlighted quotations will convey some of the overall messages regarding "the problem" and "the solutions":

- "Quality mathematics education for all students is essential for a healthy economy." (p. 1)
- "Mathematical literacy is essential as a foundation for democracy in a technological age." (p. 8)
- "Mathematics must become a pump rather than a filter in the pipeline of American education." (p. 7)
- "Mathematical illiteracy is both a personal loss and a national debt." (p. 18)
- "Mathematics offers special opportunities as a productive vocation for disabled persons." (p. 24)
- "Children *can* succeed in mathematics. If more is expected more will be achieved." (p. 2)

But there is also substantive writing in the report, and its deceptively simple style does draw on evidence, argument, and authoritative opinion. Its 114 pages contain

---

are thereby weakened. But then I don't need convincing; nor am I representative of the intended audience.

[4] The writer who drafted the report, Lynn Steen, is a man with a track record in writing articles which bring the latest mathematical research ideas to a wider readership. He is a writer with a fine sense of "audience," and *Everybody Counts* bears his stamp.

sections on such topics as human resources, curriculum, teaching, mobilizing for curricular reform, and moving into the next century.

The references section gives "chapter-and-verse" sources for those who need more detail, and its comprehensive bibliography contains 164 references. The lists of members of the National Research Council's Mathematical Sciences Education Board, the Board on Mathematical Sciences, and the Committee on the Mathematical Sciences in the Year 2000, with which the report begins, is a veritable who's who of people concerned with mathematics education, college mathematics, and research mathematics. It is a good example of its genre and worth buying for that reason alone.

Its relationship with the *Standards* report is clarified by the following paragraphs from its "Action" chapter:

> Although pressure for change is high, little consensus exists on what mathematics students ought to learn now, much less on what they will need for the future. Lack of national focus has created such disparities among standards that it is difficult to discuss curricula in meaningful and productive contexts. Teachers have received such mixed signals that even the best of them often do not know which choices to make in those few classes where they have some discretion over what to teach.

The new *Curriculum and Evaluation Standards for School Mathematics*, being published in early 1989 by the National Council of Teachers of Mathematics (NCTM), focuses national attention on specific objectives for school mathematics. That report, the draft of which has been reviewed extensively by teachers and the public, has received widespread support in the mathematical and educational communities. It represents the first effort ever to establish national expectations for school mathematics. (p. 89)

And later in the same chapter:

> Once vigorous dialogue and grass-roots actions begin forging national consensus on goals for school mathematics, several important national objectives must be addressed:
>
> - Establish new standards for school mathematics.
> - Upgrade the teaching profession.
> - Make assessment responsive to future needs.
> - Strengthen collegiate mathematics.
>
> The first of these will emerge, with sufficient effort, following public dialogue about the NCTM *Standards*. The second is currently being advocated through the work of the National Board for Professional Teaching Standards. The third, assessment, may require a new, cooperative, national mechanism to unlock the stranglehold that state and national testing programs–largely secret–have on today's classrooms. Finally, strengthening college and university mathematics–including specific attention to those who become teachers, how they teach, and what they teach–is the primary task of the National Research Council's Committee on the Mathematics Sciences in the Year 2000. (p. 95)

Everybody Counts doesn't just focus on *school* mathematics, but gives direct attention to the need for tackling three other immensely complex institutions; namely, the teaching profession, the testing industry, and the collegiate curriculum. However, I am not clear why the textbook publishing industry was omitted from this list of "controlling" institutions. Because of the size of its market, the demographic and cultural range of the communities served, and the "nationalizing" role of school textbooks, it has always seemed to me that the textbook industry is one of the most powerful conservative forces in the United States. It is hard to imagine that reform of the order intended in these documents could happen without a shake-up in that industry's practices. Perhaps it was omitted from the list because it was thought that NCTM's close connections with that industry could provide the route for appropriate discussion and negotiation. Whether negotiation of the required order could in fact happen without a strong case being made in the public document is an open question. In any case, there is plenty of work to be done concerning the three institutions that are included.

NCTM's *Standards* is also a crucial component in the political process of reform. In appearance, it is a far more "meaty" document: 272 pages, with different sections dealing with curriculum standards for grades K-4, grades 5-8, and grades 9-12, and sections on evaluation standards and "next steps."

It is apparently intended for professionals in the mathematics education field. Also available is a fifteen-page Executive Summary, which provides an overview of the general arguments and thrust of the report. *Standards* was produced by five working groups within NCTM. Reading the list of names makes clear how closely related the production of the two reports was–six members of the NCTM's thirteen-person Commission on Standards for School Mathematics were also on the NRC's Mathematical Sciences Education Board. We clearly have an impressive concerted effort, which to my knowledge is unique in the history of U.S. mathematics education.

The choice of the word *standards* is an interesting one. As is stated in the preface: "As school staffs, school districts, states, provinces, and other groups propose solutions to curricular problems and evaluation questions, these standards should be used as criteria against which their ideas can be judged" (p. v). There are, then, thirteen standards for the K-4 curriculum, thirteen for grades 5-8, fourteen for grades 9-12, and fourteen evaluation standards. In each of the sets of curriculum standards the first four have the same headings—Mathematics as Problem Solving; Mathematics as Communication; Mathematics as Reasoning; Mathematical Connections–and the remainder are more content-oriented. Some of the terminology is familiar; some will be new and challenging to teachers. So if, for example, we choose Standard 3. Mathematics as Reasoning, for grades 5-8, we find familiar phrases such as: "Recognize and apply deductive and inductive reasoning"; and more challenging ones like: "Make and evaluate mathematical conjectures and arguments" and "Validate their own thinking" (p. 81).

If we choose Standard 11, Probability, for the same grades, we find challenging notions such as: "Appreciate the power of using a probability model by comparing experimental results with mathematical expectations" and "Develop an appreciation for the pervasive use of probability in the real world" (p. 109).

It is when we read in the Introduction about the need for standards that the rhetoric appears more aggressive. Three reasons for having standards are given:

> First, standards often are used to ensure that the public is protected from shoddy products. . . .
> Second, standards often are used as a means of expressing expectations about goals. . . .
> Third, standards often are set to lead a group toward some new desired goals. . . . (p. 2)

Examples and analogies from other professions are used to support the arguments here. Then comes a fascinating paragraph:

> Standards are needed for school mathematics for all three purposes. Schools, teachers, students, and the public at large currently enjoy no protection from shoddy products. It seems reasonable that anyone developing products for use in mathematics classrooms should document how the materials are related to current conceptions of what content is important to teach and should present evidence about their effectiveness. (p. 2)

I find this section very striking. It implies that not only is there a desire to see mathematics educators upgrade their professional activities, and the report contains many ideas about that, but that the mathematics education community needs to go on the offensive against both the testing industry and educational publishers. These two institutions are apparently the ones who have developed the "shoddy products" that have restricted the aspirations, and therefore the realities, of mathematics classrooms across the nation to such depressingly low levels. NCTM is in effect saying to mathematics educators: 1) here is a comprehensive and thorough set of objectives which those *other than* teachers must also meet if teachers are to be at all successful at achieving these reforms, and 2) these objectives have a very influential body of opinion behind them. Professionals within the mathematics education community have allied themselves with the very powerful mathematicians' lobby and are prepared, as never before, to challenge the authority of arguably the two most controlling forces in mathematics education in the United States.

Moreover, besides curriculum standards, which carry with them many implications for changes in content, text materials, teaching style, and assessment, there is a separate set of evaluation standards designed to put yet more pressure on those who stand accused of marketing outmoded, irrelevant, and "shoddy" test and evaluation products. The fourteen evaluation standards include headings such as "Multiple Sources of Information," "Appropriate Assessment Methods and Uses," "Mathematical Power," and "Communication," all of which will enrich the debate about appropriate evaluation.

Standards is also amply illustrated with a wide range of examples. There are plenty of arguments, images, and evidence with which to expose "shoddy" practices. The report has claimed the moral "high ground" of mathematics education. The necessary negotiation, persuasion, and educational lobbying that must follow will be much more successful from such a position.

Having painted a broadly brushed picture of the two reports, and having also aligned myself with the idealists in the struggle for quality mathematics education, let me now probe the ideas in these reports a little further. What is the character of the quality mathematics education the two reports are offering?

The kind of excellence envisioned here is best caught by the oft-used phrase "mathematical power," coupled with the idea of equity–the mathematical empowerment of all people. As we learn in *Standards*, mathematical power "denotes an individual's abilities to explore, conjecture and reason logically, as well as the ability to use a variety of mathematical methods effectively to solve non-routine problems.... In addition, for each individual, mathematical power involves the development of personal self-confidence" (p. 5). Hence the presence of standards other than those concerning curriculum content.

This is mathematics as activity and process, not simply as mastery of content. The image of mathematical power builds on knowledge developed through grappling with student performance criteria, but goes way beyond the usual minimalist prescriptions. It does this by including performance criteria more associated with professional mathematicians in the mathematical sciences. It is certainly a powerful picture.

Is there, though, too much of an emphasis on mathematics as performance? Certainly the aim is not just to let others who are "more gifted" get on with it. We are told firmly that mathematics should not be a spectator sport, and occasionally the phrase "mathematical training" is used. Clearly the mathematics community wants more people to study mathematics for longer and to go further in the subject. The emphasis is on *doing* mathematics and on valuing the doing of mathematical activities.

Though one can certainly applaud that image, it is nevertheless reasonable to ask whether this approach, of itself, will offer all students a satisfactory and quality mathematics *education*. Mathematical activity can very easily become mathematical training and, to make an analogy, we know that physical training is very different from physical education—the criteria for performance and skill development are very different from those for judgment and the development of understanding. A mathematical education must not only encourage mathematical activity, but also offer the experiences of reflecting *about* mathematics. Also, while a mathematical training can certainly benefit those who succeed, what *educationally* does it offer those who don't? Those who will not become professional mathematicians do not need mathematical training, but they do need mathematical education which will, in a democracy, empower them to understand and ultimately to evaluate the activities of practicing mathematicians in all walks of life. Competence without a reflective perspective is no education.

Moreover, nowhere in these reports is there stated an *educational* ideal of which mathematics education is a part or to which it can contribute. Nowhere is there serious argument on the relationship between mathematics and other subjects. Nowhere is there discussion of the possibility that decline in enrollments is a choice in favor of other concerns, rather than away from mathematics. The view being expressed here is not only that more people *could* practice mathematics for longer if only we trained them better, but that they *should* and must for their own good and for the good of the nation. Given the political urgency associated with other concerns, like the polluted environment, global coexistence, and social inequalities, this proposal could seem in some people's eyes rather like trying to increase enrollments in the chess class on the Titanic.

Three aspects of the new vision could be particularly significant, however, provided that they are realized. The first concerns the relatively straightforward idea that the K-12 mathematics curriculum content should be radically overhauled. It has been a source of concern to me and other foreign educators to see how much the curricular content has stagnated over the two decades since I taught in the United States. Of course, in the 1960s the curricular content was dominated by New Math language and protocols, but at least that challenged the students. The "deadening" curriculum effect of minimal competency allied to the conservativism of the textbook industry is sometimes difficult for a foreigner to appreciate, but I cannot help feeling that the content proposed in *Standards* is not just new (although comparable to what is now being taught in other countries), but will come as a distinct shock to some people. For example, the use of computer graphics utilities for solving equations and inequalities, computer-based methods of successive approximation, three-dimensional geometry, matrices, functions, probability and statistics, the use of scientific calculators — all are expected in the three-year core course proposed for *all* high school students.

As the authors of *Standards* themselves admit: "Initially it may appear that an excessive amount of curriculum content is described in the 9-12 standards." They then go on, rather optimistically one feels, to explain that: "When this course is evaluated, however, it should be remembered that the proposed 5-8 curriculum will enable students to enter high school with substantial gains in their conceptual and procedural understandings" (p. 125).

These ideas are certainly being taken on throughout Europe and Australia. The recent developments in mathematics education worldwide — the search for more creative, investigational work, the release from routine procedures through the use of calculators, the possibilities offered by computers, the calls for greater cultural and historical awareness in mathematics teaching, or more use of discussion and argument in classrooms — can all be found under the umbrella term of "mathematical empowerment," and they all exist in practice.

The second new aspect, which I find particularly interesting, is the fourth curriculum standard in each group, entitled "Mathematical Connections." This is a very different standard from any of the others. To meet it, "the curriculum should include deliberate attempts, through specific instructional activities, to connect ideas and procedures, both among different mathematical topics and with other content areas (p. 11). Thus for the middle grades there should be opportunities to "apply mathematical thinking and modeling to solve problems that arise in other disciplines, such as art, music, psychology, science, and business" and opportunities to "value the role of mathematics in our culture and society" (p. 84). I personally would have been happier with the verb "reflect on" instead of "value" in that last statement, in line with my preference for a little more education and a little less training. But having said that, attempts to get an idea like "Mathematical Connections" on the developmental agenda will surely begin to develop a broad mathematical education.

The third aspect concerns the proposed changes in instructional practice. When one compares the many journal descriptions of current classroom practice with researched developments and innovations, a wide gap emerges. My reading of this section of *Standards* and the others on instruction suggests that many of the proposals *do* accord with best practice ideas in the United States and other countries, and do not support a narrow "training" approach to instruction. I am extremely heartened when I read, for example, that increased attention will be given to "the active involvement of the pupil in constructing and applying mathematical ideas," "the use of a variety of instructional formats," and "the use of calculators and computers as tools for learning and doing mathematics" (p. 129). There is certainly little I would quibble with here, and I feel that the document makes a good case, with examples, for the proposed changes in instructional practice. Whether or not teachers will agree with my assessment remains to be seen, since implementing these proposals will require many changes in the majority of U.S. classrooms.

What will need developing, therefore, is initial and inservice teacher education. There will be a great deal of work to be done in this area given the demands that *Standards* makes in both curriculum and instruction. With a comprehensive agenda of the kind laid out in these two reports, it is going to be necessary to develop a "critical mass" of educators who will be able to generate a national momentum for teacher education in mathematics. Documenting and providing the resources for initiatives will be an important activity, as will mobilizing the educational research community to support the reform developments with appropriate research activity. Significant teacher education initiatives are happening in the world and it will be important for the mathematics teacher-education community in the United States to build on these. I sense that this community is not as cohesive or powerful as it could be, but it must surely see that there is everything to be gained from throwing itself wholeheartedly into the *Standards* reform agenda, and a great deal to be lost by not doing so.

Let us therefore turn to the question of the prospects for the proposed reforms. Can these reforms be realized? Is the system capable of delivering this agenda? To begin to answer these questions one must be aware that there are teachers, schools, curricula, and texts which do satisfy the standards. There are plenty of examples of innovative and imaginative teaching at all levels — from kindergarten to college level. The best of the United States can rank with the best anywhere, and *Standards* illustrates and documents the possibilities.

It is therefore going to be necessary to recognize and support those individuals, teachers, and institutions that clearly demonstrate the feasibility of the ideal. *Everybody Counts* does recognize the pioneering work of Jaime Escalante in teaching his urban Mexican-American students. It also highlights the fact that one of the most brilliant mathematical minds of this century exists in the handicapped body of Cambridge's Stephen Hawking, and points out that mathematics offers many opportunities to people with various disabilities. Models are crucial to reform — the "existence hypothesis" always needs establishing–and these will need to be encouraged at the national, regional, and local levels. One danger of using the language of "crisis" and "disaster" is that those who are currently exemplifying better practice may not feel that their efforts are being appropriately recognized.

What is important to support is the authority for any innovative work, and this is where the reports can be particularly useful. Our own Cockcroft report in the United Kingdom, entitled *Mathematics Counts*, was a particularly useful document for showing those in powerful positions who needed convincing that one's innovations were respectable and responsible, rather than way-out or outrageous.[5] Both reports more than satisfy that standard – the lists of names and institutions have already been referred to. These documents could be helpful in supporting already existing initiatives and developments. To that extent the reforms can be delivered, given appropriate support.

Let me now speculate a little further on whether the reforms will be realized. Undoubtedly something good will come from the work that was put into producing these reports. There must be a strong desire for change of this sort to happen if the reformers have got as far as they have. The mathematics education community apparently realizes that it now has the best chance it is ever likely to have to move away from the "back-to-basics" ideology. There is apparently so much at stake that the community does seem prepared to invest the necessary human and institutional resources to turn the system around. What, though, of those schools and colleges where little is happening of an innovative and empowering nature? This of course is the problem area – particularly as this reform agenda is not going to make a reluctant teacher's life any easier. It is a very challenging agenda, and teachers at all levels are going to need every possible ounce of encouragement, structuring, and support if anything substantial is to be achieved.

What are the constraints to be overcome? The United States always seems to me to be high on leadership but low on followership, and bodies like NCTM and the NRC are going to have to do much more than just point to the new direction. Some tough bargaining and negotiating will need to take place with textbook publishers, state adoption boards, teacher training institutions, and assessment and testing boards if the conservative restraints on development are to be loosened. The multiplicity of school districts and the nature of local educational control makes the whole concept of national reform both questionable and extremely difficult at the local level. The first key constraint therefore concerns how much political energy the mathematics education community can stimulate, nourish, and sustain. In the United Kingdom, the process was helped by the significant involvement of many people and groups before the publication of the Cockcroft report. The key factor here will be whether the prior engagement of active groups has been sufficient to ensure their energy and support in the post-report phase of the reform process. The politics of educational change in a democratic society requires involvement, not imposition.

Another key constraint is financial, and, coming from a country where money for education has always been in short supply, I don't necessarily believe that this reform program is a particularly expensive one. However, I am enough of a realist to know that any educational reform in the United States takes a lot of money. Once

---

[5] Wilfred H. Cockcroft, *Mathematics Counts* (London: Her Majesty's Stationery Office, 1986).

again, to judge by the names and credits in *Everybody Counts*, the mathematics community has indeed got its connections well established and could tap into the necessary financial sources. Nevertheless, as in the United Kingdom, there has been considerable pruning of the U.S. budget for education during the 1980s. This could either mean that nothing will happen in the 1990s, or that the "Future of the Nation" argument in these reports will stimulate the government and the business community to put up the necessary financing for reform. Given the current political and economic turmoil in Europe, there could well be some additional funds invested in these kinds of initiatives.

The status of the reformers is also an important factor in the struggle for the acceptance of any reform. In this case, two influential bodies have combined their efforts and, particularly striking to me, neither NCTM nor NRC appears to have a reputation for reforming zeal. Both are normally fairly conservative bodies, the NRC perhaps by choice and the NCTM more because of its size. We have learnt in the United Kingdom that conservatives can certainly stimulate reform.

On the other hand, it is perhaps a little surprising that there is not much reference to the research literature concerning mathematics learning and teaching. There is no impression of the existence of a substantial body of research on which, for example, the proposals in *Standards* are based. Recommendations and exhortations appear to be supported only by opinion–authoritative opinion, it is granted–but opinion nonetheless. It is, however, going to be necessary to mobilize *all* the supportive forces if the reforms are to be realized, and I would anticipate a need for some detailed research back-up to the prescriptive statements. Already the research community in mathematics education has sensed the need, but their involvement has come too late in the reform process to have much impact on the kinds of reforms being proposed.[6]

However, what is required is research support that can be given by exemplifying, documenting, and analyzing different approaches to successful intervention practice at any level. The increasing use of ethnographic and case study methods could be bonus in this regard. Perhaps, also, we will begin to see less of a research emphasis on fine-grained psychological inquiry in favor of more sociologically informed studies of change and development within educational institutions. Educational institutions are the most neglected and necessary of the social aspects of mathematical education, and should be a research priority.

There is scant reference in these reports to the social context and constraints controlling the present practice and reality of mathematics education in the United States. The problems "in the field" are complex and real: social and racial inequalities, lack of adequate funding, teenage dropouts, and distressingly improverished educational and learning environments. I am not convinced that the vision painted by these two reports will make a great deal of immediate sense to some of the

---

[6] Research Advisory Committee of the NCTM, "NCTM Curriculum and Evaluation Standards for School Mathematics: Responses from the Research Community," *Journal for Research in Mathematics Education, 19 (1988)*, 338–344.

professionals struggling to cope under increasingly frightful conditions. To paraphrase a point made by Dan Lortie in his classic work *Schoolteacher*, teachers have a built-in resistance to change if it doesn't directly affect what they perceive to be the constraints on their work situation.[7] Education is about goals and ideals, but schools and educational institutions are places of work and are about rules, which have their own existence and morality.

Whether any reform proposals in mathematics teaching will figure highly in many teachers' current agendas is doubtful. Whether mathematics teaching can improve, on its own, and without any other educational factors changing, is extremely unlikely. In an atmosphere of low academic expectations for both teachers and students, where morale is also low, where the day-to-day grind of compulsory schooling creates its stresses and burn-outs, there is little reforming zeal to nurture. While I can understand the reluctance of the *Standards* authors to grapple with these aspects (unlike the authors of *Everybody Counts*, which does present a stark image of reality at times), in my view, there should have been much more attention to the teachers' reality, the mathematics department's reality, and the institution's reality.

To read *Standards* one would think that the *only* interaction that is important for a learner's mathematical development (and therefore for the nation's mathematical development) is that which takes place between teacher and learner. Of course I am not denying the importance of this interaction, but I do feel that it is irresponsible to ignore completely the social context in which it occurs. Students come to the educational encounter with their own and their community's cultural and social value systems, and the teachers are employed in an educational institution with its hierarchies, social and physical constraints, administrative procedures, internal politics, and community pressures. The lived reality for the participants in the pedagogical interaction needs to be addressed in any educational reform process. It is ignored in *Standards*; indeed, the whole sociopolitical dimension of mathematics education is never referred to. The model is of autonomous teachers operating in their own private classrooms with their own agendas and their own students. This model is a myth, however, and reformers who believe in its validity are deluding themselves.

What is going to be needed is at least another document which addresses directly the different aspects of the politics of reform in mathematics education. Otherwise the brave ideas in *Standards* will be mentally placed on the shelf alongside all the other recommendations which have never been followed through. It is not my task here to say what such a document should contain; but a section, for example, on "how to get change going in your school" might help. It could also list standards for school provision or for mathematics department facilities. It is well known, for example, that the acquisition of computers and other high-tech equipment in sufficient quantities necessitates room space and technician back-up. Mathematics teachers with that support can easily argue that they can no longer teach in any old classroom in the school, and a suite of mathematics rooms becomes a necessity. This space can act as a focus for increasing the profile of mathematics in the school,

---

[7] Dan Lortie, *Schoolteacher* (Chicago: University of Chicago Press, 1975).

with the use of appropriate resources, display, and so on. The important point is that part of the political process necessary to improve mathematics teaching is for those who are *not* teachers to press for better working conditions for those who are. More resources and human support are essential components of improved working conditions. But that is only one small example of the political aspects that still need to be addressed.

Whether the mathematics curriculum at large figures highly in many teachers' agendas is also doubtful. For the elementary teacher, mathematics (or rather arithmetic) is only one of several subject concerns. For the middle grade teacher, life among adolescents has its own peculiar dynamic. High school teachers' thoughts are dominated by their particular courses, and college and university teachers have to concern themselves with research and publications. As *Everybody Counts* says, "Mathematicians seldom teach what they think about – and rarely think deeply about what they teach" (p. 40). How do the reformers propose to change this state of affairs?

On this same point, it is important to say that although at present most of the focus of reform lies with public schools' mathematics teaching and curriculum, *Everybody Counts* makes a very strong case for putting college-level mathematics under the spotlight:

> To improve mathematics education, we must restore integrity to undergraduate mathematics. This challenge provides a great opportunity. With approximately 50 percent of school teachers leaving every seven years, it is feasible to make significant changes in the way school mathematics is taught simply by transforming undergraduate mathematics to reflect the new expectations for mathematics. Undergraduate mathematics is the bridge between research and schools and holds the power of reform in mathematics education. (p. 41)

Much of the success of reform, then, will hinge on the college-level mathematics community's ability to rethink, to innovate, and to recognize its central role in this movement. My sense is that this community is far less used to the spotlight than are high school teachers. I can already hear the defensive cries of college and university teachers, backed up, of course, by thoroughly argued and plausible reasons as to why they should continue to do what they have always done. "Independence" is not a luxury open to high school teachers, and in most countries that quality is jealously guarded by college and university teachers. But independence also carries with it the need for responsibility, and we will have to see more public attention being paid to that value if the aspirations of *Everybody Counts* are to become a reality.

Finally, there is the "international," or perhaps xenophobic, dimension. Where Sputnik stimulated the New Math movement, the challenge of foreign commercialism is perhaps now fueling the Mathematical Empowerment movement. Just as the reports address the educational problems within the country, so there is also a widespread recognition of foreign economic competition. While it is hard to imagine that intelligent researchers and educators can believe the tenuous relationship described earlier between school mathematical achievement and national economic performance, both communities of scholars who produced these two reports are certainly aware of developments in mathematics education in other countries. The "minimal competency" doctrine was essentially a U.S. phenomenon – while most

U.S. educators were coping with all of its manifestations, the rest of the mathematics education world was moving on, albeit with half an eye on the "back-to-basics" political pressures.

Educational systems everywhere have been upgrading their curricula and teaching approaches. New materials and technology have been making their mark, and national reports and books about new ideas in mathematics education have been appearing with increasing frequency. As international conferences have become firmly established on the mathematics education scene, U.S. colleagues have become concerned that the system of mathematics education in the United States is getting left behind.

Not only has the world of mathematics teaching moved on, so has the world of mathematics learning. The results of the Second International Mathematics Survey from the International Project for the Evaluation of Educational Achievement (IEA) are gradually appearing and they do not make pleasant reading for the U.S. colleagues (nor for us in the United Kingdom.).[8] The superior mathematical performance of East Asian students in different comparisons appears to have made a striking impact on U.S. morale, although it is interesting to learn that Japanese mathematics educators are also concerned about their students' (relatively) weak performance on test items concerned with non-routine problem solving.

Sensibly, in my view, *Everybody Counts*'s response to the external "situations" is to argue for developing a truly American approach to mathematical education rather than merely emulating what is happening elsewhere. This in a sense is what Americans have always done, but there is more overt "Americanism" in this report than in most reports on mathematics education: "Imitating others is no solution. The United States must find a strategy that builds on the tradition of this country, one whose strength lies in this nation's unique tradition of local initiative and decentralized authority" (p. 90). Grand words, but one could certainly make a case that it is "this nation's unique tradition of local initiative and decentralized authority" that has produced precisely the situation now facing the reformers. More worrying is that the *Standards* report doesn't offer any strategy that will enable the constraints of decentralized authority to be overcome.

Ultimately, this call to respond to the challenge from international competition will be the significant factor determining this reform's acceptability. Much will depend on the kinds of arguments and political influence used to get change happening on the ground. The reports have created an analysis of the problem, a vision of the future, and a challenge to the mathematics education community and, more important, to everyone else with a controlling stake in that community's activities. The analysis is reasonable, the vision *is* a significant one, equal to the "excellence with equity" criterion that it proposes. What remains to be seen is whether the challenges can be met. They deserve to be, but I have many doubts. I shall continue to watch and support if I can, with interest.

---

[8] See, for example, F. Joe Crosswhite et al., *Second International Mathematics Study: Detailed Report for the United States* (Champaign, IL: Stipes Publishing, 1986).

Chapter 12
Mathematical Power as Political Power – The Politics of Mathematics Education

Christine Keitel and Renuka Vithal

Focusing Social and Political Aspects of Mathematics as a Necessary Part of Research

In 2000, at the 9th International Congress of Mathematics Education (ICME 9) in the Discussion Group on Social Aspects of Mathematics Education, Alan Bishop stated that the current mathematics education situation in most countries is non-democratic, that the curriculum is a mechanism of governmental control rather than educational enlightenment, that the commercial textbook 'business' community controls the materials available to teachers to an extent that teachers are slaves of the textbook rather than autonomous and enlightened users, and that assessment is still primarily and predominantly a mechanism for selecting the mathematical elite only. These statements also showed his disappointment with developments in reforms he had considered as revolutionary before[1] (Bishop 1990a). The few promising developments in mathematics education he saw in the increasing availability of personal technology, in the growth of the vocational education sector, the growth of the informal sector through web-access, and – this was his big hope – in the increasing professionalisation of mathematics teacher associations, and the growth of equal and fair, collaborative and politically sensitive research within the mathematics education community.

Within this broad arena, Alan Bishop's seminal contribution to mathematics education has been the exploding of the myth that mathematics is culture-free and value-neutral. In his key work *Mathematical Enculturation: A cultural perspective on mathematics education*, he writes:

C. Keitel
Fachbereich Erziehungswissenschaft und Psychologie, Freie Universität Berlin, Habelschwerdter Allee 45, 14195 Berlin, Germany
e-mail: keitel@zedat.fu-berlin.de

[1] "Mathematical Power to the People" (Bishop 1990a) is the rather euphoric title of his review of the USA publications "Everybody counts" of MSEB & CMS (1989) and "Curriculum and Evaluation Standards" of NCTM (1989).

P. Clarkson, N. Presmeg (eds.), *Critical Issues in Mathematics Education*,

> My aim is to create a new conception of Mathematics which both recognizes and demonstrates its relationship with culture – the notion of mathematics as a cultural product, the environmental and societal activities, which stimulate mathematical concepts, the cultural values which mathematics embodies – indeed the whole cultural genesis of mathematical ideas ... (and the) many implications for mathematics education ... which concern the mathematics curriculum, the teaching process and teacher preparation. (Bishop, 1988a, pp. xi–xii)

He advanced these ideas in the mathematics education community when, as one of the most influential and long-term editors of the journal *Educational Studies in Mathematics*, he contributed to the early mainstreaming of "Mathematics Education and Culture"[2] through a special issue which explored a broad set of concerns captured above as well as political issues of mathematics education. No doubt, any history of mathematics education research and practice will recognize the 80s as the era of the rise of the cultural dimensions of mathematics and mathematics education. These developments were emphasized as social processes but by the end of that decade culminated in a greater, more explicit focus on the political aspects. D'Ambrosio's plenary address in 1984 in ICME 5 laid the "Socio-cultural bases for Mathematical Education" and established Ethnomathematics as a field of study and practice (D'Ambrosio, 1985). By ICME 6 in 1988 a special day added to the program and captured in a special UNESCO edition of the proceeding "*Mathematics, Education and Society*" with more than 90 papers on such issues (Keitel, Bishop, Damerow, & Gerdes, 1989); and Mellin Olsen's (1987) "The Politics of Mathematics Education" revolutionized the dominant understandings and framings for the teaching and learning of mathematics.

Alan Bishop emphasized that the challenge for all of us within mathematics education is how to support developments from within our own, often highly politically dominated educational contexts, and with our own limited resources to mainly focus on the following:

- Developing social, cultural and political dimensions of mathematics education academically and as a serious field of empirical investigation and theoretical study;
- Analyzing and questioning associations, organizations, and conferences of mathematics education socially and politically.

These two goals characterize a major concern of Alan Bishop with policy issues in general and cultural issues in particular. He has identified policy studies as an important gap in mathematics education and has argued for stronger links between research, policy, cultural practices and their underpinning values. His main concerns aim at investigations on how we could further develop and ensure the autonomy and professionalisation of mathematics educators and teachers and their active role in their associations or organizations, although this raises many contradictory proposals. Alan Bishop has provocatively put forward questions such as: "*Do we speak for ourselves and are we really autonomous in our professional and*

[2] Educational Studies in Mathematics, 1988b, Vol. 19, No. 2

organizational decisions? Who are those driving policy in mathematics education and research, and for whom do they speak?"

These questions have recently become actual in Germany. 2008 is announced as the "Year of Mathematics" and mathematicians are organising a large number of conferences, workshops, research activities and popular papers to convey a positive image and celebrate their science, in newspapers, in television spots and huge advertisements, with a so far incomparable energy and intention to popularize and value mathematics and gain money for research. To persuade the masses, the slogans propagate predominantly the "fact" that mathematics is the purest science and does not convey any political intention and is independent of policy statements or political activity, it is just very important research for any future. To persuade the political and economical stakeholders, they promise that mathematics is useful for political goals. With different means, mathematicians want to convince ordinary people as well as politicians to support financially much more mathematical research through large funding, claiming that mathematical research as such is necessary and a social demand. The contradiction between a propagated image of mathematics as "policy- and value-neutral" and an obvious policy behind mathematics research projects is demonstrated bluntly and without any doubt. A research cluster in mathematics that gained the prize of excellence in a nation-wide competition, which is mathematically based and strictly application oriented research, was sold as "mathematics research as a service for key technologies", and quite frankly claims that the most and overall importance of mathematics is the fact that it can be used for actual and important political goals. The question is: has mathematics as a science that is taught and as an area of research for which money is needed ever been policy-free and politically independent?

In South Africa fundamental curriculum transformation in the last decade has seen the new national mathematics curriculum statements explicitly refer to mathematics as a cultural product, creating a space to engage mathematics critically, socially, historically and as part of a process of recognizing indigenous knowledge systems. Yet in practice and in the media, mathematics performance in general and especially the high stakes national end-of-schooling grade 12 examination as well as international comparative studies, there is little evidence of taking account of the wide social and cultural diversity of learners and teachers or indeed even the still existing large disparities in basic conditions of schooling. Racial categories, which have been reintroduced and continue to be used to assess redress of the disadvantage of South Africa's apartheid past, show that "Black African" learners, the majority of whom continue also to be poor and are taught in poorer classroom settings, are consistently found to dominate the lowest scores. Although a range and complex of factors explain this result, little reference is made to the curriculum policy itself, its values and ideological underpinnings and the contradictory practices of assessments, language of instruction, and contexts for learning which proceed as if mathematics and mathematics education is a-political, culture- and value-free. Contradictory public naming and shaming of poor mathematical performance and at the same time celebration of isolated cases of exceptional performance by poor schools, which are often the result of heroic efforts or some specific opportunity that cannot be replicated at a system level (e.g. access to nearby higher education institutions or considerable extra tuition) continue with little focus on the quality, processes and content of learning shaped by the well known more recent debates of *ethnomathematics, realistic mathematics, critical mathematics* or *social-concerns* of constructivism. Mathematics education research appears in this case to have spoken to policy but has failed to connect or speak to practice, and failed to influence and inform a public understanding of mathematics and mathematics education that changes itself.

The following issues and concerns address what Alan Bishop has raised again and again in his contributions to mathematics education research, theory, policy and practice and what should serve as a guideline for the discussion:

- Social, cultural and political views about mathematics and mathematics education: Is mathematics really for All?
- Social justice and mathematics education: dream-team or nightmare?
- Social and political conflicts by poverty, violence and instability: How has mathematics education participated in or contributed against these?
- Challenges and perils of globalization in international collaboration: Who gets a fair deal?
- Contradictory demands and measures for new qualities of mathematics and mathematics education research: Who determines what concerns mathematics education and research?

Mathematics as a Means of Political Power

Mathematics is perceived today as one of the most powerful social means for planning, optimizing, steering, representing and communicating social affairs created by mankind. And by the development of modern information and communication technologies based on mathematics, this social impact of mathematics came to full political power: Mathematics is now universally used in all domains of society, and there is nearly no political decision-making process in which mathematics is not used as the "rational" argument and the "objective base" that is considered as replacing political judgments and power relations: mathematics as objective truth and free of politics.

For the ordinary citizen today, in contrast to public political propaganda, it becomes increasingly difficult and sometimes impossible to follow developments of mathematics, in particular mathematical applications with information communication technologies (ICT), and to evaluate their social use appropriately, because specialization and segmentation of mathematical applications are often extremely hard to understand. The principal insight into their necessity and a basic acknowledgement of their importance in general are often confronted by a complete lack of knowledge of concrete examples of their impact, not to speak about alternatives. However, competencies to evaluate mathematical applications and ICT, and the possible usefulness or its problematic effects, now seem to be a necessary precondition for any political executive and for real democratic participation of citizens. The new challenge is to determine what kind of knowledge and meta-knowledge in a mathematized society is needed and how to gain the necessary constituents. This has relevance for both those in power or government as well as those outside. Not understanding the mathematical basis of decisions and not having the means for understanding how mathematics participates in decision making by the citizenry has different kinds of consequences in societies with different levels of literacy. For example political consequences such as protests about the lack of, or unfair distribution of state resources such as education, housing, health and so on create politically vested interests in producing a mathematically competent and literate populace who can be convinced by rationality and mathematically grounded arguments. But what it exactly means to be mathematically competent or literate has been changing over time.

The development of competencies for decision-making under conditions that include the coding and processing of knowledge by means of systems of symbols (e.g. book-keeping, planning models, calculation of investment or pensions, quality control, theories of risks, IT in banking etc.) and their complete mathematisation, is not only an actual problem. History offers numerous examples of the fact that similar problems arose at various times, although in historically specific forms and with historically specific solutions. In particular, examples of historically radical breaks in the organization of decision-making could be referred to, i.e. examples of developments, in which the mutual effect of changes of knowledge systems and innovations in the technology of using information have questioned traditional mechanisms of decision-making, as well in the educational and social policy as in policy for scientific development, which in the long term have been replaced by new forms.

Studies in the history and philosophy of sciences show that changes in the forms of symbolizing and processing of information usually had mutual effects on social organizations and led to new structures of scientific knowledge. Together with the change of structures of knowledge, different characteristic styles of thinking and in particular, worldviews were developed, which also effected fundamental changes in the process of political decision-making in general social goals, and in the means and measures to pursue them, accompanied by changes in the allocation of resources in a society. Who has access to these knowledge systems and can critique the impact of their applications within the state and in civil society, has consequences for the functioning of democracies.

Since the beginning of social organization, social knowledge of exposing, exchanging, storing and controlling information in either ritualized or symbolized (formalized) ways was needed, therefore developed and used, and in particular information that is closely related to production, distribution and exchange of goods and organization of labor (Nissen, Damerow, & Englund, 1990). This is assumed as one of the origins of mathematics. Early concepts of number and number operations, concepts of time and space, have been invented as means for governance and administration in response to social needs. Mathematics served early on as a distinctive tool for problem solving in social practices and as means of social power. Control of these social practices and the transmission of the necessary knowledge to the responsible agents were often secured by direct participation in social activities and direct oral communication among the members. Ritualized procedures of storing and using information have been developed since Neolithic revolution, during the transition to agriculture and permanent living sites, which for example, demanded planning the cycles of the year.

The urban revolution and the existence of stratified societies with a strong division of labor induced symbolical storage and control of social practices by information systems based on mathematics, which were bound to domain-specific systems of symbols with conventional meaning: Mathematics employed as a technique and a useful and necessary tool; and the governing class disposed of mathematics as an additional instrument of securing and extending its political power and authority.

A new, most consequential perspective of mathematics emerged in Ancient Greece. For the Greeks, mathematics was detached from the needs of managing ordinary daily life and from the necessity of gaining their living. Instead, through scientific search for fundamental, clearly hierarchically ordered bases, creating connections and elements of a systemic characterisation of existing formal mathematical problem solving techniques and devices, and independent of any specific practical intention, they reformulated mathematics as a scientific system and philosophy. A (Platonic) ideal theory was to be further discovered and constructed by human theoretical thinking and reasoning, not by doing or solving practical problems. Being able to think mathematically was a sign of those who had political power. By viewing mathematics as the structure underlying the construction of the cosmos and number as the basis of the universe and emphasizing a hermetical character of the mathematical community, the ground was laid for the high esteem of mathematics as a segregation means of political and social power.

Over the centuries, the traces of structuring the world by human rational activity became more numerous, appropriate for dealing with it, and more imposing. There were several fields in which the mathematisation of the real world and of social life advanced more remarkably; among these notably are architecture, military development in both fortification and armament, the mining industry, milling and water-regulation, surveying, and before all, manufacturing and trade (Damerow & Lefévre, 1981). The extension of trade from local business to long distance exchanges prompted the emergence of banking. For the functioning of this activity an unambiguous form of clear and universal regulation was needed: the system of book-keeping was invented. In none of these systems were mathematical devices without political intention and power relations. Mathematical structures determined whole fields of social practices, which are working until the present day (Damerow, Elwitz, Keitel, & Zimmer, 1974).

This phenomenon becomes especially visible in colonial history. Colonization ensured an early "internationalization and globalization" of these systems and secured mathematics a place within. For Alan Bishop (1990b), mathematics, "as one of the most powerful weapons in the imposition of Western culture" (p. 51) participated in "the process of cultural invasion in colonized countries" through at least three agents: trade and commerce e.g., units, numbers, currency; mechanisms of administration and government e.g., through number and computation systems needed to keep track of people; and imported systems of education e.g., mathematical curricula for the elite few needed to politically administer colonies.

For a subject seen as most outside the influence and realm of the social or cultural, mathematics is deeply implicated in the distribution and enactment of political power. Described nowadays as both a "gate-keeping" and a "gateway" subject, high mathematical performance distributes opportunity to access high status and high paying professions and positions.

Social Needs for Mathematics Education in the Course of History

On the base of the mathematical and scientific developments, the achievements of the 15–19th century entailed an explosion of trades, crafts, manufacturing and industrial activities with an impressive diversity, ingenuity, and craftsmanship developed and required in numerous professions (Renn, 2002). This period also coincides with a period when Europe "discovered new worlds" of the South and East and confronted different cultures, knowledge, skills and economic traditions The ability of a greater part of the population to deal appropriately with fundamental systems of symbols like writing and calculating became a precondition for the functioning of societies: Elementary (mathematics) education and training – although clearly restricted to defined needs – was established as reaction to social demands, either prior or during vocational training and various professional practices. Parallel to upcoming educational institutions and in concert with them, mass production for unlimited reproduction of knowledge enabled and asked for standardisation and canonical bodies and representations of knowledge. A reflection and restructuring of existing knowledge on a higher level was demanded: Meta-knowledge had to be developed that offered standards of knowledge and their canonical representations for educational purposes; at the same time meta-knowledge as orienting knowledge became an immanent condition for developing new systems of knowledge, in particular for sciences like mathematics that were perceived to a greater part as independent of immediate practical purposes.

In the 19th century, the competition between the bigger European states, inspired by a strong and fateful ideology of national superiority and ambition, drew attention to "knowledge as power", making public education a central interest of governments. Industrialisation was accompanied by an increasing autonomy of systems of scientific and practical knowledge. To be a mathematician, somebody who does mathematics and nothing else, emerged as a new profession. Mathematics was conceived as an autonomous subject domain, without immediate practical use in other domains, and mathematicians worked as scholars at university or as high-level schoolteachers. These school and university systems were also replicated, even if in smaller versions, in colonized states together with their knowledge domains and curriculum content and organization.

As in sciences, in social services and administration, specialization and professionalisation of experts were requirements in all branches of economy. Constructing and creating new knowledge became *a* precondition for the material reproduction of society, not a consequence. Specialization was a condition for creating new knowledge, but at the same time bore the risk of disintegrating more comprising systems of knowledge and making integration in a wider context difficult. Partial knowledge must be generalized and incorporated on a level of meta-knowledge.

Material and intellectual resources for this new knowledge project were generated from within states but could also be appropriated from outside. In order to maintain superiority and power (political, intellectual, economic), colonizing countries valued and incorporated any knowledge from their colonies if it could be absorbed into existing canons, and subordinated or dismissed that which could not. Ethnomathematics, as a recent movement and case in point, recognizes the construction of this specialized meta-knowledge, its conflictual and collaborative history, and its location and rootedness within societal contexts, but its important role and place in mathematics and mathematics education is not equally valued in all countries or contexts.

Mathematics Education as a Public Need and Task

In the 19th century in many countries, public and state controlled two partite school systems were created – higher education as mind formed for an elite, elementary education to transmit skills and working behavior for the majority, the future working class. Not only has this dual provision of education become institutionalized over time, mathematics has come to play an important role in this stratification. At least one observable consequence is that mathematics has become implicated by its fragmentation into an abstract mathematics for the few and a kind of so-called numeracy or arithmetic for the rest.

W. v. Humboldt's notion of "Bildung" comprised learning as very general and independent of any special matter or subject area, as universal as possible with strong emphasis on general themes in the humanities: philosophy, history, literature, art, music, but also with an emphasis on pure mathematics and sciences. The ideal was the completely cultivated, best educated human being, and "Bildung" was not a process, an attitude and a path as much as an accomplishment. And that was to be conveyed by means of public education, at schools and at universities. Mathematics became a subject in higher education institutions for the elite and governing class because of its formal educational qualities, e.g., educating the mind independently of a direct utilitarian perspective, and fostering general attitudes to support the scientific and science-driven technological development. This makes a political dimension of mathematics visible within continued contradictory claims of mathematics as neutral and value-free. As university education reforms take place, mathematical knowledge, regarded as foundational, participates in the international and globalised trade on the knowledge market as very much central to the new knowledge economy that feeds and grows the commerce in science and technology.

In the elementary or general school for workers and farmers, only arithmetic teaching in a utilitarian sense was offered to secure the necessary skills for the labor force and to secure acceptance of formal rules and formal procedures set up by others. The restriction was clear – mathematics education for the few was strictly separated from skills training for the majority. This corresponded to a separation of mathematics education as an art and science in contrast to mathematics education as a technique; scientific knowledge and conceptual thinking versus technical, algorithmic, machine-like acting. Despite efforts to integrate mathematics education by integrating both aspects and focusing on awareness of the differences, in all segregated or differentiated types of schooling, these divisions determine schooling systems and differentiate also school curricula in discriminatory ways.

Advances in science and technology in the twentieth century have sharply impacted on schooling and society and opened for questioning of the kind of public mathematics education that is needed socially. It has had implications, on the one hand, for the "producers" of "pure" mathematics as an abstract knowledge, and on the other hand, for the "users" of outcomes or products of mathematics. The science and technology dominated era has brought with it different kinds of consumers of mathematical knowledge – those who apply mathematics as a technical knowledge to create technological tools of all kinds and those who just use those tools (often as a black box) as consumers (Skovsmose, 2007). The provision of mathematics education as a public need and task has to deal with this complexity within a context of unequal access to these very mathematical facilities. The question of who will be the producers and who will be the different kinds of users of mathematics becomes important because mathematical power, through a science and technology driven knowledge and skills market economy then participates in a particular distribution of other kinds of (political or economic) power. Large numbers of people also get left completely outside mathematics and its many products made available through science and technology – the poor, the homeless, those caught in conflicts and wars, those who do not get access to education and so on.

Mathematics for All? A Failure of Social Ambition

In the 20th century mathematics became the driving force for almost all scientific and technological developments. Mathematical and scientific models and their transformation into technology had a large impact not only on natural and social sciences and economics but also on all activities in social, professional and daily life. This impact increased rapidly through the development of the new information and communication technologies based on mathematics, which radically changed the social organization of labor and our perceptions of knowledge or technique on a global scale to an extent that is not yet fully explored or understood. Mathematics education became necessary and a precondition for social and economic development. But still there were two different aspects attributed to mathematics and mathematical knowledge that asked for complementary contradictory conceptions of mathematics education.

On the one side, mathematics as a human activity in a social environment is determined by social structures, hence it is not interest-free or politically neutral. On the other side, the continuous application of mathematical models, viewed as universal problem solving procedures, provide not only descriptions and predictions of social actions, but also prescriptions. The increasing social use of mathematics makes mathematical methods and ways of argumentation into quasi-natural social rules and constraints. This usage creates a social order based on mathematical criteria, and becomes effective in social organizations, hierarchical institutions like bureaucracy, administration, management of production and distribution, institutions of law and the military, etc. Social and political decisions are turned into facts, constraints or prescriptions for individual and collective human behavior then follow. Mathematics education has had to respond to the unfolding of these processes and this may be observed across countries despite differences in resources, different cultures and social, political and economic systems, histories and education provisions and literacy levels.

New perspectives of the social role of (mathematical) knowledge and general education only partly gained political acceptance and support: "Mathematics education for all", "Numeracy", and "Mathematical Literacy". These concepts were differently substantiated and received different interpretations and supporters. The New Math movement had started to introduce mathematics for all through a formally unified, universally applicable body of theoretical knowledge of modern mathematics exposed to all, but had to be revisited and discarded as a solution. Intensive work in curriculum development created a wide range of different and more and more comprehensive approaches combining new research results in related disciplines like psychology, sociology, and education and developed this vision further (Howson, Keitel, & Kilpatrick, 1981; Sierpinska & Kilpatrick, 1998). A variety of conceptions promised to describe the socially necessary knowledge in a more substantiated form and to integrate scientific mathematical practices and common vocational or professional practices and their craft knowledge, or conceptual and procedural knowledge, or mathematical modeling and application. One example of this was the highly celebrated "Everybody counts"- a document from USA, at first valued very much by Alan Bishop (Bishop, 1990a).

However, the most radical development within and outside of mathematics as a discipline, was caused by the invention of electronic media and the new possibility of data-processing and control. The immediate consequence, based on the integration of human-mental and sensory-information processing techniques within machines, is the creation of technologies which take over human information processes and independently determine social organization. This new development is called globalization of knowledge. The technological integration of new representation forms and the distribution of knowledge in a global net of knowledge, represents the greatest challenge for a restructuring of political power and decision making processes about the way in which information is gained and used, is available to anybody everywhere with access to the internet, and is useful for individual interests on a the basis of the person's mathematical and scientific literacy.

Information and communication technologies are the foundation for communication which is an essential aspect of globalization: access to and exchange of information and knowledge from anywhere in the world, quickly and cheaply. On the one hand, that leads to a general acknowledgement of cultural diversity, but on the other hand also to universalisation and domination by certain languages and cultural positions – e.g. the English language and Euro-American or Western belief systems, encompassing a variety of knowledge traditions and knowledge systems. The pervasiveness of this development can be seen in how information communication technologies have seeped across communities, nations, cultures and jumped poverty-wealth and urban-rural divides within and across countries.

The social role and impact of mathematics has dramatically changed throuh the development of modern information technologies based on mathematics. Mathematics is ascribed a new utility value, which has never before been so strongly indubitable as it is now. Illustrative examples for new technological and the most effective applications of mathematical methods are numerous, e.g., computer-based simulations are applied in many different areas like modeling of climate changes, motor vehicle crash tests, chemical reaction kinetics through building process-oriented technical machines, dynamical system models in macroeconomics and biology.

Software packages allow the most complex calculation processes for many applications in forms of black boxes, like statistical processes in quality control, research on market and products, risk theories for porte-feuille-management in assurance companies, computer based algebra systems and software for modeling in sciences and engineering. Mathematics, as the basis of many technologies, is effective although only invisibly. In this way it is a theoretical base for formal language in informatics, as foundation of coding algorithms for industrial robots or in the daily-used scanners, mobile phones, cash corners or electronic cashiers. New technologies in return feed back with great impact on mathematics as a discipline itself. Besides traditional applied mathematics, new directions combine applied sciences with experimental procedures like techno-mathematics, industrial mathematics, and theory of algorithms.

New procedures in some application areas are celebrated not only as new means to ends or a refined methodological repertoire, but furthermore, as a new paradigm. In contrast to classical applied mathematics, which was oriented towards and restricted to the representation of mathematical structures of a reality existing completely independently of any subjective intention, new forms of applications do not hide the fact that interests and intentions always guide the construction of a model, as well as specific goals and convictions.

Indigenous knowledge systems of technologically and scientifically less advanced or slower progressing countries are now smoothly recruited into this movement and global growth but only if the mathematical knowledge can be directly linked into the system of abstraction or application. If not, it remains outside mainstream developments, which explains in part, the difficulty for ethnomathematics, which gives value to local knowledge, to impact formal curricula in any serious way.

Mathematics and information technology not only provide descriptions and explanations of existing reality, but they also *create* new reality. As a basis of social technologies like arithmetical models for election modes, taxation models, calculation of interests and investment, calculation of costs and pensions etc., mathematical models are transformed into reality, establish and institutionalise a new kind of reality. This process can be reconstructed and analysed as the development of implicit mathematics. Patterns of social acting and formal structures are transformed via formal languages into algorithms or mathematical models which can be rectified and objectified as social technologies (Davis & Hersh, 1986; Davis, 1989; Keitel, Kotzmann & Skovsmose, 1993).

Such results of applications of mathematics are often encountered in communication situations mainly shaped by conflicting interests where they serve to justify opinions and to stabilize attitudes. Graphical representations of information, for example, are excellently structured, provide sufficient overview and relative universality of readability, but are also appropriate means for accentuation and guiding perceptions into the wrong directions. In such communication processes the possibilities for interaction between interpreters are usually restricted. Even neglecting the fact that credibility often depends on the prestige of the participants, the prestige of mathematics as such often serves to suggest objectivity and objective goals and intentions. Thus the regulation and democratic control of actual and future research, development and application processes of mathematics and mathematics education demand a specific competence and knowledge as a basis for decision making on the part of the politicians and new knowledge for evaluation and democratic control on the part of the citizens.

In countries where this kind of knowledge and skills base is inadequate or weak in mathematics and scientific communities, among the politicians and policy makers and in the overall literacy levels of the general population there are serious consequences for the development of those countries: for the growth of their science and technology fields, which are central for meeting the demands for basic provisions of education, health, housing; and for the quality of their systems of governance and political stability; as well as for their participation in the global economies, research and political positioning.

Mathematical Literacy for Political Power and Critical Citizenship

The pervasiveness of economic thinking and interests have successively created so high a pressure of economic orientation that educational aims and the subject matter are marginalized unless they provide, justification in terms of economic interest (Woodrow, 2003). New notions like "mathematical proficiency, competency or literacy", "educational standards" and "benchmarks" are expressions of such economic interests. They are a major concern of politicians but also a pressure for educational researchers and practitioners. They are the key issues in the recent political debates and disputes about mathematics education, which broadened after the release of international comparative studies like the Third (later called Trends in) International Mathematics and Science Study (TIMSS) and the Programme for International

Student Assessment (PISA) and their ranking of test results. Early debates dominating mathematics education literature that linked notions of mathematical capacities with concerns for citizens to participate meaningfully in society and specifically in a state's decision-making have given way to narrow concerns with test performance scores and been justified with tenuous links to a country's economic position and performance.

Proclaiming that the PISA tests are based on "definitions of mathematical literacy" that are underpinned by fundamental and widely accepted educational research results, and that it is absolutely unproblematic to test these kinds of competencies or proficiencies on a global scale and to rank countries' performances, produced strange and urgent political measures to be taken in some of the countries that did not perform well, called for by the alarmed public and the media. Results of tests like PISA are used as a reference and base for decisions in educational policy and interventions, in particular in cases when they show that only a small part of the tested students or adults have reached a higher level of competencies in the international comparison. In some countries these have led to a regime of ongoing national testing, importation of learning materials and copying classroom pedagogies, often with little attention to the broader societal conditions, cultures, educational histories and values that produced a particular (high or low) performance.

PISA claims to measure in its test of Mathematical Literacy, those competencies of young adolescents that enable them to participate in democratic decision-making processes:

> Mathematical Literacy is the capacity to identify, to understand and to engage in mathematics and make well-founded judgments about the role that mathematics plays, as needed for an individual's current and future life, occupational life, social life with peers and relatives, and life as a constructive, concerned and reflective citizen. (OECD, 2000, p. 50)

As this definition clearly demonstrates, each attempt to define Mathematical Literacy is confronted with the problem that this cannot be done exclusively in terms of mathematical knowledge – to understand mathematised contexts or mathematical applications and to competently use mathematics in contexts goes beyond mathematical knowledge. One of the first research studies to explore such cross-curricular competencies, by investigating the ways in which mathematics is used in a social–political practice, had unexpected and surprising results (Damerow et al., 1974).

Conflicting conceptions of Mathematical Literacy are numerous, although the conflict is not always recognized. Jablonka (2003) analyzed what research on Mathematical Literacy can and cannot do. She showed that different perspectives on Mathematical Literacy vary considerably with the values and rationales of the different stakeholders who promote them. The central argument underlying each of her investigations is that it is not possible to promote a conception of Mathematical Literacy without at the same time – implicitly or explicitly – promoting a particular social practice of mathematics, be it the practice of mathematicians, of scientists, of economists, of professional practices outside science and mathematics, etc. She argues that Mathematical Literacy focusing on citizenship in particular, refers to the possibility or the need to critically evaluate the most important issues of the surrounding society or culture of the students – a society and culture that is itself very much shaped by practices involving mathematics. In her conclusion, she emphasizes that the ability to understand and to evaluate different practices of mathematics and the underlying values has to be a component of Mathematical Literacy.

Mathematical Literacy must be understood as functional in relation to pedagogical postulates. But by reducing the concept of Mathematical Literacy to the descriptions of the process of its measurement cannot be justified, while conclusions of these comparisons are formulated mostly in terms of daily language or connected with highly demanding and complex meanings and connotations of the concepts. This also begs the question of mathematics teachers' education, willingness and/or ability to engage and teach these complex, broader and demanding notions and concepts of any Mathematical Literacy curriculum.

The demands and threats of a Knowledge Society are referred to in most political declarations and justifications for educational policy. From an international or global point of view, this includes investigating what approaches towards knowledge perceptions are taken in different countries, at the levels of policy and of practice; what are the most important knowledge conflicts at various social levels, and in particular in the educational systems, e.g., clashes between students' personal knowledge and the knowledge presented by teachers, between knowledge systems, between 'modern/popular' cultures and traditional cultures, between teachers' and students' views (Clarke, Keitel, & Shimizu, 2006); and on the more general level, for example how are global technologies – especially the World-Wide-Web, television and print media – used to promote or diminish diversity, or what effects of inequality are reproduced.

The question of how mathematics is perceived and used in political debates and decision-making processes, in particular in decisions that concern mathematics education, is a necessary complement to be studied. Case studies to investigate which connections are established between results of comparative studies on mathematical competencies and the attributions of causes and effects deduced from them in the public debate are a very promising beginning, although diverse political interests shape very much the research results. To collect and analyze which criteria for political decisions and which forms of decision-making processes are defined and stated, which kind of controlling mechanisms to secure quality is foreseen or used and on what the credibility of results is based, in particular in the media, is greatly needed. To reconstruct the origin and history of such studies and confront criteria and decisions for selecting the participating institutions and experts, then contrast the official publications of national and international projects and the reconstruction of views and conceptions held by participating experts in interviews would be a necessary amendment of those studies. The history of the social reception of these studies is to be interpreted in the light of conflicts of interests and different interest groups. An analysis of published statements of all stakeholders in political decision-making processes and of representatives of interest groups in industry and economy has been started, complemented by interviews with teachers, mathematicians and experts in the ICT-area. The interpretations of these statements in the light of the factual political interests have to be re-analyzed on the basis of the historical accounts of mathematics as means of social power and of the actual account of modern mathematics as a scientific discipline and technology provider. This is what Alan Bishop asked for in many of his papers and it adds hopefully to a broader and more substantially defined conception of becoming mathematically literate, showing clearly the close relationship between mathematics and policy and leading hopefully to a debate about what and how much mathematics is needed to educate or create politically aware, well informed and critical citizens for a democratic society.

Mathematics Education Research on the Margins of Political Power

While there has been a focus on the relation between research and practice in mathematics education, it is in the nexus of research, policy and practice that mathematics education research interfaces political power and to date remains relatively under-explored. It may be argued that mathematics education research has not substantially impacted on policy and therefore failed to substantially shape curriculum and classrooms practice. A complex web of stakeholders and connections explain this lack, but at least two reasons may be advanced. One is the failure of mathematics educators to engage key policy makers or analysts and politicians within state systems and to do the kind of research, writing and theorizing that could speak to those in government and administration decision making; and the second is the rise of international testing regimes that have quickly filled the gap left by mathematics educators, who may dominate mathematics education scholarship in conferences and journals, but do not substantially impact the typically deeply contested political terrain of policy.

The first may be ascribed in part to the lack of attention mainstream mathematics educators have paid to policy studies and to macro system level practice questions as areas for inquiry and investigation. It is only in the last decade that this gap in the literature has been identified as an important silence. Although mathematicians and mathematics educators have engaged critique of policy and reforms that have taken place, as Bishop (2002a) observed, there is a lack of journals and conferences that explicitly deal with interactions between policy makers, researchers and practitioners. Lerman, Xu and Tsatsaroni (2002) confirmed this lack in their analysis of articles published in one of the mainstream and longstanding mathematics education journals, *Educational Studies in Mathematics*. They reported that in more than a decade, since 1990, they found only four articles that addressed policy issues and in a way that spoke to both researchers and policy-makers. Some attempts at addressing this lacuna are beginning to emerge. The *Second International Handbook of Mathematics Education* (Bishop, Clements, Keitel, Kilpatrick, & Leung, 2003) carried a section on the policy dimensions of mathematics in which a major political aspect – the relationship between mathematics and economic development is explored (Woodrow, 2003). The questions of how research agendas are constructed, in whose interests, and who are the audience for mathematics education research is especially pertinent.

This lack of focus on the policy dimension means that mathematics educators have not understood or substantively impacted macro and system level issues, and this gap has been quickly taken up by those doing large scale performance testing, especially those engaged in the numerous recent international studies run by companies in public-private partnership contracts that are mostly profit-oriented only (Jahnke & Meyerhöfer, 2007). Moreover, researchers involved in such studies, who are often not well versed in mathematics education research and concerned in the main with evaluation of education systems, have also succeeded in dominating media attention and thereby shaped public understanding and perceptions about mathematics education and explanations for performance. Politicians, who eventually strongly influence curriculum policies and resource deployment, are more responsive to media critique due to their concerns with public image and opinions and also because the outcomes of these studies are presented in short "sound bites" or "bottom lines" and rather simplistic terms in a discourse that they can relate to or are willing to engage. As already alluded to, in many countries a testing regime has taken hold as evidenced in widespread and continuous mathematics testing being done at various grade levels despite serious problems of methodology and questions of validity, and with little regard for inequities and complexities of context variation, within a single country and across countries.

Within this situation mathematics and mathematics education related associations and organizations have an important role in interfacing with policy makers and politicians as emphasized by Alan Bishop in many of his publications. Even if mathematics educators are recruited into government, what has to be recognized is that government policy environments and bureaucracies have their own discourses which may be difficult to challenge, change or influence. For mathematics education researchers and practitioners, the task, on the one hand, is that of developing a domain of understanding that produces the knowledge, skills and spaces to engage the political spheres of government; and on the other hand, to be able to interact with media as "public intellectuals" to influence public opinion and understanding of mathematics and mathematics education related issues that are not reductionist and do not narrowly attribute blame only to teachers and learners for particular performances, especially when such findings are presented in competitive league-tables type formats.

Mathematical Classroom Pedagogy and Political Power

Globally, widespread mathematical curriculum reforms have taken place or are taking place in many countries. This phenomenon has created further impetus to research a broad range of mathematics education related practice issues at a system or policy level nationally and internationally. The two justifications often observed among politicians and policy-makers for their respective mathematics curriculum reforms are those linked to economic gains – as already mentioned – to be achieved from greater proficiency in mathematics; and those linked to enhancing democracy by improvements in their citizens' mathematical literacy. This rhetoric of governments that seek to link (mathematics) education to democracy, are often included in the preamble to their curriculum policies and refer typically to notions of critical citizenship. Though not always visible in the mathematics content or topics of an official or intended curriculum, there is often some suggestion or indication about what is expected in a mathematics classroom. But a critical citizenship that is to be authentically developed through classroom pedagogy puts the development of any real and viable mathematical power in direct conversation with political power. In any critical education, not only is this dialogue brought to the centre of the teaching-learning situation, it carries with it an uncertain and risky environment in which the outcomes are seldom predictable.

A teacher, who by definition of her profession is middle class – even if from a poor or working class background – in a class of learners from poverty contexts and who chooses to engage a pedagogy for social justice and evokes the power of mathematics to raise awareness and knowledge about their life conditions, cannot control that the very same learners may act against her or the school. In contexts in which learners sit together in a classroom from very diverse cultures, with deeply conflicting values and unequal conditions, who come to confront the evidence, arguments, and justifications provided by a mathematics intending to reveal and work through these differences, embodied in the very learners in the classroom, will invariably engage a pedagogy of conflict and dialogue (Vithal, 2003). Mathematics classroom pedagogies that open spaces for confronting inequities and the social, cultural, political dimensions of a mathematical problem, project or topic, even if they competently engage other disciplines and knowledge domains, have to contend with conflicts. Mathematical power recruited as political power in practice brings a range of conflicts – historical, social, cultural, political, economic – into the classroom. It is the recognition of cultural conflicts, for example, that brought a shift in Alan Bishop's work from a focus on mathematical enculturation (1988a, 1988b) to acculturation (Bishop, 1994, 2002b), which may be interpreted as a changing concern from how learners are inducted in a mathematical culture to how a mathematical culture itself transforms through its interaction with the diversity of cultures of learners.

If the argument that societies of the 21st century are fraught with different conflicts is accepted then it follows that so are schools, which are microcosms of society (Skovsmose, 1994). Mathematics classrooms that open for teaching-learning approaches that authentically connect mathematics to context would have to engage such conflicts through dialogue (Vithal, 2003). But any pedagogy of conflict and dialogue has to take account of and be able to deal with yet another aspect that has not adequately featured in mathematics critical or social justice pedagogies – the anger that accompanies knowledge and awareness of injustices and the pain that accompanies the bringing of suffering into the classroom and the opening of wounds. Different relations of power, embodied in the conflicts, can manifest in different ways and would need to be managed and engaged in different classrooms, for example, if all learners are from a similar cultural, gender or class background to one in which the divides of poverty and wealth or associations made with "perpetrators" or "victims" are attached to the very learners or even to the teacher in that same classroom.

It is in this respect that mathematical power being deployed as political power within contexts of conflict (be they historical, cultural, economic or social) needs to engage yet another or additional dimension of pedagogy, a pedagogy of forgiveness (Waghid, 2005). A mathematics that reveals inequities and injustices of the past or present often produces feelings of hate and resentment. In such contexts, Waghid (2005) notes that "learning about forgiveness can become useful in enhancing pedagogical relation" (p. 226) and that when teachers and learners "cultivate forgiveness" it becomes a way to "engender possibilities whereby people are attentive to one another" and can engage "imaginative action" to move forward. Such pedagogy requires the creation of spaces, in the first instance, for "truth" to be told so that reconciliation can occur. Only then can dignity be reclaimed, compassion shown and respect and friendship built. Critical, feminist, and social justice mathematics pedagogies seek to mobilize the power of mathematics knowledge and skills to such overt political and social agendas. But in order for restoration and peace to emerge, such pedagogies will have to attend to more than official mathematical knowledge in the situation.

However in this educational setting, drawing from South Africa's Truth and Reconciliation Commission (1998), there are multiple "truths" and not only a "mathematical truth" that needs to be engaged for reconciliation to occur, which may also be relevant to any such pedagogy. The Commission identified a conceptual framework comprising four kinds of truth. The first was factual or forensic truth based on "objective information and evidence" (p. 111), which may be analogous to a canonical mathematics knowledge or the official curriculum. The second, personal or narrative truth in which "everyone should be given a chance to say his or her truth" (p. 112) resonates with ideas in (socio) constructivism in which the mathematical knowledge of each individual is recognized and respected. Third, social or dialogue truth, "the truth of experience established through interaction, discussion and debate" (p. 113) resembles the propositions of Ethnomathematics which argue for recognition of the mathematics of different social and cultural groups in society from the past and in the present. Fourth, a healing and restorative truth as the "the kind of truth that places facts and what they mean in the context of human relationships" (p. 114), which could be translated to suggest that the different mathematical knowledge and skills of each individual and group has to be brought into dialogue, and has to be connected and contextualised.

Within this framework, "Mathematical truths" may be deemed one kind of truth within a framework of "truths" that need to get expression in a mathematical classroom. It also alludes to how conflicts and dialogues that take place in such classrooms would need to be handled if mathematics education is not only about increasing knowledge and awareness of inequities and injustices but also to be a means for forgiveness and healing. Often mathematics is presented as a single and only truth, the most objective or neutral and this one truth that is most valued while community or personal mathematical knowledge, skills and practices are subordinated or silenced. Mathematical literacy, as has been shown earlier, explicitly calls for broadening learning to integrate context, but often fails to recognise what this means within diverse unequal contexts and the capacity and competence needed by teachers to work through "truth conflicts" that may be present in a mathematics classroom. In this, what counts as "mathematical truth" and mathematics education itself is being challenged and seen as undergoing change.

Alan Bishop (1994), already early on, pointed to how mathematical knowledge and mathematical practices of traditional or indigenous cultures, non-Western societies, and different groups within each society are being increasingly recognized and incorporated into curriculum policies, practices and texts, as conflicts of language, as mathematical concepts and procedures, symbolic representation, attitudes, values, beliefs and cognitive preferences that are the subject of ongoing debates within and across countries. No doubt much more research and reflection is needed about the politics of mathematics pedagogies and classrooms and the outcomes for learners if mathematical power is to be realized as political power.

References

Bishop, A. J. (1988a). *Mathematical enculturation: A cultural perspective on mathematics education*. Dordrecht: D. Reidel Publishing Company.

Bishop, A. J. (Ed.) (1988b). Mathematics education and culture. Special Issue, *Educational Studies in Mathematics, 19*(2).

Bishop, A. J. (1990a). Mathematical power to the people. Review of "Everybody counts: A report to the nation on the future of mathematics education by MSEB and Committee on Mathematical sciences in the Year 2000" and "Curriculum and Evaluation Standards for School Mathematics" by NCTM, 1989. *Harvard Educational Review, 60*(3), 357–369.

Bishop, A. J. (1990b). Western Mathematics: The secret weapon of cultural imperialism. In: *Race & Class, 32*(2), 51–65.

Bishop, A. J. (1994). Cultural conflicts in mathematics education: Developing a research agenda. *For the Learning of Mathematics, 14*(2), 15–18.

Bishop, A. J. (2002a). Research, policy and practice: The case of values. In P. Valero, & O. Skovsmose (Eds.), *Proceedings of the Third International Mathematics Education and Society Conference*, 2–7 April, (pp. 227–233). Helsingor, Denmark: Centre for Research in Learning Mathematics.

Bishop, A. J. (2002b). Mathematical acculturation, cultural conflict, and transition (pp. 193–212). In G. de Abreu, A. J. Bishop & N. C. Presmeg (Eds.), *Transitions between contexts of mathematical practices*. Dordrecht: Kluwer Academic Publishers.

Bishop, A. J., Clements, M. A., Keitel, C., Kilpatrick, J., & Leung, F. (Eds.) (2003). *Second international handbook of mathematics education*. Dordrecht: Kluwer Academic Publishers.

Clarke, D., Keitel, C., & Shimizu, Y. (Eds.) (2006). *Mathematics classrooms in 12 countries: The insiders' perspectives*. LPS-series No. 1, Rotterdam: Sense Publishers.

D'Ambrosio, U. (1984). Socio-cultural bases for Mathematical Education. In M. Carss (Ed.), *Proceedings of the fifth international congress on mathematical education*, (pp. 1–6). Adelaide & Boston: Birkhausen.

D'Ambrosio, U. (1985). Ethnomathematics and its place in the history and pedagogy of mathematics, *For the Learning of Mathematics, 5*(1), 44–48.

Damerow, P., Elwitz, U., Keitel, C., & Zimmer, J. (1974). *Elementarmathematik: Lernen für die Praxis? Ein exemplarischer Versuch der Bestimmung fachübergreifender Curriculumziele* [Elementary mathematics: Learning for practice? An exemplary attempt to determine cross-curricular goals in mathematics]. Stuttgart: Klett.

Damerow, P., & Lefévre, W. (Eds.) (1981). *Rechenstein, Experiment, Sprache. Historische Fallstudien zur Entstehung der exakten Wissenschaften*. [Historical case studies on the origin of exact sciences]. Stuttgart: Klett-Cotta.

Davis, P. (1989). Applied mathematics as social contract. In C. Keitel, et. al. (Eds.) *Mathematics, education and society*, UNESCO Document Series No. 35, Paris: UNESCO, 24–28.

Davis, P. J. & Hersh, R. (1986). *The mathematical experince*. Boston: Houghton Mifflin.

Howson, G.A., Keitel, C., & Kilpatrick, J. (1981). *Curriculum development in mathematics*. Cambridge: Cambridge University Press.

Jablonka, E. (2003). Mathematical literacy (77–104). In: Bishop, A., Clements, K., Keitel, C., Kilpatrick, J., & Leung, F. (Eds.), *Second international handbook of mathematics education*. Dordrecht: Kluwer Academic Publishers.

Jahnke, T., & Meyerhöfer, W. (Eds.) (2007). *PISA & Co. – Kritik eines Programs*. [PISA and Company – a critique of a program]. Berlin: Franzbecker.

Keitel, C., Bishop, A. J., Damerow, P., & Gerdes, P. (Eds.) (1989). *Mathematics, education and society*. Science and Technology Education, Document Series No 35, Division of Science Technical and Environmental Education, Paris: Unesco.

Keitel, C., & Ruthven, K. (Eds.) (1993). *Learning from computers: Mathematics education and technology*. Berlin: Springer.

Keitel, C., Kotzmann, E., & Skovsmose, O. (1993). Beyond the tunnel vision: Analysing the relationship between mathematics education, society and technology (242–279). In C. Keitel, & K. Ruthven (Eds.), *Learning from computers: Mathematics education and technology*. Berlin: Springer.

Lerman, S., Xu, G., & Tsatsaroni, A. (2002). Developing theories of mathematics education research: The ESM story. *Educational Studies in Mathematics, 51*(1–2), 23–40.

Mellin-Olsen, S. (1987). *The politics of mathematics education*, Dordrecht: D. Reidel Publishing Company.

Nissen, H., Damerow, P., & Englund, R. K. (1990). *Frühe Schrift und Techniken der Wirtschaftsverwaltung im Vorderen Orient: Informationsspeicherung und - verarbeitung vor 5000 Jahren* [Early writing and technologies of the management and administration of economics in the Near East: Information storing and processing 5000 years ago]. Bad Salzdetfurth, Germany: Franzbecker.

OECD (Eds.) (2000). *Programme for International Student Assessment* (PISA). Paris: OECD

Renn, J. (2002). *Wissenschaft als Lebensorientierung – eine Erfolgsgeschichte?* [Science as orienting life – a story of success?] Preprint 224, Berlin: Max-Planck-Institut für Wissenschaftsgeschichte.

Sierpinska, A., & Kilpatrick, J. (Eds.) (1998). *Mathematics education as a research domain: A search for identity*. Dordrecht: Kluwer Academic Publishers.

Skovsmose, O. (1994). *Toward a critical philosophy of mathematics education*. Dordrecht: Kluwer Academic Publishers.

Skovsmose, O. (2007) Mathematical literacy and globalization (pp. 3–18). In W. Atweh, B. Barton, A. C. Borba, M. Gough, N. Keitel, C. Vistro-Yu, & R. Vithal (Eds.), *Internationalisation and globalisation in mathematics and science education*. Dordrecht: Springer.

Truth and Reconciliation Commission of South Africa (1998). *Report*, Volume One. Cape Town: Juta & Co., Ltd.

Vithal, R. (2003). *In search of a pedagogy of conflict and dialogue for mathematics education*. Dordrecht: Kluwer Academic Publishers.

Waghid, Y. (2005). Education, imagination and forgiveness. *Journal of Education, 37*, 225–241.

Woodrow, D. (2003). Mathematics, mathematics education and economic conditions (11–32). In A. J. Bishop, M. A. Clements, C. Keitel, J. Kilpatrick, & F. Leung (Eds.), *Second international handbook of mathematics education*. Dordrecht: Kluwer Academic Publishers.

Personal Notes

Our collaboration with Alan Bishop goes back to each of our first steps into mathematics education.

Christine Keitel's collaboration with Alan Bishop goes back to her first steps into mathematics education in the early 1970s. She had worked as a mathematician in areas of modern mathematics during her studies, became curious about research in mathematics education when the hype of New Math in USA and Great Britain reached Germany, and therefore wanted to engage critically in surveying and analysing their advantages and failures. At the newly founded University of Bielefeld, the federal-wide research institute for mathematics education IDM, which had started to search for academic wisdom and acceptance for an ambitious program on research in mathematics education (didactique), she worked on curriculum analysis of the recent New Math development in USA, GB and Germany. She had a number of wonderful chances to meet and discuss with international colleagues who were already well known and acknowledged who came to conferences in Bielefeld to act as advisors, path finders, critiques for research directions and to provide the necessary international orientation and critique for the institution and its plans or programs. Among them, of course, was Alan Bishop. To meet Alan was inspirational and challenging. He influenced her whole professional life by always engaging in most exciting discussions, exchanging papers from very early on, but also by an enlightening collaboration in projects and the writing of books. He always offered challenging and encouraging critiques for plans and projects. Alan trusted Christine to take up various professional roles: review editor for ESM (Educational Studies in Mathematics) under his editorship and one of the chief organisers of the 5th Day Special Program "Mathematics, Education and Society" of ICME 6 in Budapest, as well as chief editor of the proceedings published by UNESCO. Very important for her professional thinking and development was his invitation to become a member of BACOMET – a self-organised international working group of mathematics educators concerned with "Basic Components of Mathematics Education for Teachers": BACOMET was an experience of equal collaboration and challenging debates on specially crucial and promising themes for mathematics teacher education. As Alan proposed her as director of one of the projects, "Mathematics education and technology" she got the chance and challenge to clarify her long term analysis of the social role of mathematics and mathematics education with colleagues and to transform them within a broader context into a book (Keitel & Ruthven, 1993).

Renuka Vithal has had a long association with Alan Bishop spanning almost two decades and he has influenced much of her work and writing in that time. She first met him in 1990 when he lectured and supervised her thesis (Ethnomathematics and its implications for curriculum in South Africa) during her MPhil studies at Cambridge University in England. It was a time of considerable political activity and debate in mathematics education, particularly for a newly emerging democratic South Africa, and his encouragement and inspiration in continuing research in the social, political and cultural aspects of mathematics education must be acknowledged and is greatly appreciated. Alan Bishop made it possible for many emerging researchers like Renuka, especially from "developing world" contexts, to gain entry into the mainstream mathematics education, in publications, conferences, etc. and made a special effort to create and support networks for such individuals. Equally important, he also made it possible to bring the concerns of those contexts from the margin into the centre. This is evident for example when he was the editor of the journal *Educational Studies in Mathematics*, and continues into the present in his leadership of the Kluwer (now Springer) Series and his many other professional activities. It was through his significant mentorship that she succeeded in submitting her doctoral thesis for review and eventually having it published as a book *Towards a Pedagogy of Conflict and Dialogue in Mathematics Education* in 2003. More recently she has continued this co-working in having jointly edited a special guest issue of the journal *Pythagoras* (2006, No. 64) on Mathematical Literacy in South Africa. Alan Bishop has created a considerable legacy in mathematics education, which is honoured, recognized and richly deserved through this publication.

Section VI
Teachers and Research

Already in his 1976 article (Section 2), Bishop had noted that teachers do develop their own theories about education, and in particular about their own teaching. The implication is that, although these theories are not always fully articulated, sometimes not even to themselves, such theories nevertheless are very influential for their proponents.

In the key chapter for this section some of those early thoughts come to the fore. Bishop emphasises the usual imbalance of power between teachers and researchers, and pleads on behalf of the teachers for researchers to make the move to work alongside their colleagues. Other concerns that had been growing through the years are also invoked in this chapter. Issues of politics are embedded in this discussion. His interests in how the cultural and social influences on teachers and teaching break open the notion that the classroom is an insular site of action for the teaching-learning dynamic are also seen.

It is not a surprise that in this chapter the interplay of a number of Bishop's interests is also present. This chapter originated with a summary paper that Bishop was asked to deliver at the 1994 ICMI Study Conference on *What is research in mathematics education and what are its results*? (see chapter 14, this volume, for details). Hence it inevitably draws together various influences at play for the teacher. It is a useful reminder that for researchers, although we might at times concentrate in particular projects or writing assignments on specific issues, influences, and such like, in reality as researchers live out their professional lives, there is rarely such a neat compartmentalisation of the research enterprise. Threads are forever being woven to make a whole, at least for the individual researcher. In this chapter one gets a glimpse of this woven fabric on which Bishop is at work.

The two chapters that reflect on this key chapter of Bishop are quite different, but insightful. Kilpatrick takes a very reflective stance, not only situating the chapter written by Bishop, but giving interesting details of Bishop's thinking on research and teachers from the early 1970s when they first met, and from then on to the 1994 conference. Ruthven, who had worked with Bishop for several years, and who took over supervision of the mathematics education program at Cambridge University when Bishop left for Australia in 1992, looks at the developments of mathematics education in England from about that time, through a political lens. He uses notions contained in the key article, and other Bishop references, to give us a feel that many crucial ideas articulated by Bishop played a role one way or another in understanding what has happened in England.

An additional Bishop reference pertinent to this issue:

Bishop, A. J. (1992). International perspectives on research in mathematics education. In D. Grouws (Ed.), *Handbook of research on mathematics teaching and learning* (pp. 169–189). McMillan: New York.

Chapter 13
Research, Effectiveness, and the Practitioners' World

Alan J. Bishop

Introduction

The ICMI Study Conference on Research was a watershed event with a great deal of significant interaction between the participants. It was an energizing and involving experience, but as one of my tasks was to 'summarize' at the end of the conference, I tried to take a more objective stance during my involvement. I reported in my summary that I could detect certain emphases in the discourses together with some important silences. Here are some of them:

Emphases	*Silences*
analyses	syntheses
critiques	consensus-building
talking to ourselves	awareness of other audiences
political arguing (to persuade)	researched arguing (to convince)
individual cases	over-arching structures
local theory	global theory
well-articulated differences	well-articulated similarities
disagreements	agreements

It seems as if the researcher's training encourages one to analyze, to look for holes in arguments, to offer alternative viewpoints, to challenge and so forth. Or the pattern could reflect the fact that the idea of the conference itself and of this ICMI Study was seen as a challenge to the participants' authority. Certainly the 'politics of knowledge' was alive and well in all its manifestations.

So what is the concern of this chapter? Am I seeking just a nice, warm, collaborative engagement, and feeling that it is a pity that we seem to disagree so much? I have to admit that with so much conflict in the world today I do wish that there was more obvious peaceful collaboration. I also believe that with the whole idea of research into education under attack from certain ignorant politicians and bureaucrats, those who engage in research should at least collaborate more, and should spend less time 'attacking' each other.

My real concern, however, is with what I see as researchers' difficulties of relating ideas from research with the practice of teaching and learning mathematics. In

Sierpinska, A. and Kilpatrick, J. Mathematics Education as a Research Domain: A Search for Identity 33–45.

the discussion of this ICMI Study at the Eighth International Congress on Mathematical Education in Seville, many people spoke about the dangers of researchers just talking to each other, and thereby ignoring the practical concerns of teachers. Moreover, with the general tendency towards greater accountability in education, we find increasing pressures for more *effective* modes of mathematics education. This pressure needs to be responded to by the research community, and my concern is that, in general, it is not.

This chapter, then, is concerned with the researchers' relationship with the practitioners' world, and it starts by considering the issues of 'effectiveness'.

Effectiveness: The Pressure and the Responses

Mathematics has increasingly become a significant part of every young person's school curriculum, and as mathematics has been growing in importance, so has the public pressure to make mathematics teaching as effective as possible. 'Mathematics for all' (see, e.g., Damerow et al. 1986) has become a catch phrase which has driven educators in many countries to constantly review their mathematics curricula and their teaching procedures. The challenge of trying to teach *mathematics* to all students, rather than, say, arithmetic, has been one element, the other being that of teaching *all* students regardless of possible disadvantage, handicap or obstacle.

The pressure for greater effectiveness in mathematics teaching now comes both from the application of business-oriented approaches in education, supported by the theory of economic rationalism, and also from the increasingly politicized nature of educational decision making, driven by the economic challenges faced by all industrialized countries since the last world war. Education is now seen as an expensive consumer of national funds, and concerns over the perceived quality of mathematical competence in the post-school population have fueled the pressure.

Research has also contributed (perhaps inadvertently) to this pressure, firstly by supplying a range of evidence on competence, and secondly by raising expectations about the potential for achievements throughout the whole school population. It has also contributed to the concern for effectiveness by helping to generate alternative approaches in mathematics teaching. There are now many possible permutations of teaching styles, instructional aids, grouping arrangements, print media and so forth, all available within the modern classroom and school.

Add to these the possible curricular contents and emphases, curricular sequences, and examinations and assessment modes, which themselves have been stimulated by research, and the variety of *potential* mathematics educational experiences becomes bewildering, not just to teachers. Moreover, these only refer to formal mathematics education – one could continue to contemplate the further large number of possibilities if informal and non-formal educational experiences were included, along with the World Wide Web.

In the face of such a plethora of educational alternatives, the demand for knowledge about effectiveness is easily understandable, and needs to be taken seriously.

However, there is little evidence that researchers are addressing the issue seriously enough.

The editor of the recent *Handbook of Research on Mathematics Teaching and Learning* (Grouws 1992) advised:

> The primary audience for the handbook consists of mathematics education researchers and others doing scholarly work in mathematics education. This group includes college and university faculty, graduate students, investigators in research and development centers, and staff members at federal, state, and local agencies that conduct and use research within the discipline of mathematics.... Chapter authors were not directed to write specifically for curriculum developers, staff development coordinators, and teachers. The book should, however, be useful to all three groups as they set policy and make decisions about curriculum and instruction for mathematics education in schools. (p. ix)

Indeed, nowhere in that volume are the issues of effectiveness specifically and systematically addressed, although the chapters by Bishop (1992, ch. 28) and Davis (1992, ch. 29) do intersect with that domain of concern. Bishop reflects on the critical relationship between teacher and researcher, and between researcher and the educational system. Davis argues also for greater interaction between the education system and researchers.

The lack of relationship between mathematics education research and practice is documented by many references in the literature (e.g., Brophy 1986; Crosswhite 1987; Freudenthal 1983; Kilpatrick 1981). There are, however, some signs that research and researchers are relating more closely to the ideas of reform in mathematics teaching (Grouws, Cooney & Jones 1989; Research Advisory Committee 1990, 1993).

The pressure for effectiveness in mathematics teaching supports and encourages the 'reform' goal of research. In this sense, effectiveness is achieved by changing mathematics teaching since the present practice is assumed to be relatively ineffective. Thus the often quoted dichotomy between research and practice is becoming refocused onto issues concerning the role of research in changing and reforming practice. It is tempting to see 'reform' as being merely 'change' in line with certain criteria, but that analysis loses the essential dynamic associated with a reform process.

Researchers and Practitioners

There are several key issues to be faced here. Can research reform practice? Should that be its role? Clearly there are other sources for knowledge with which to inform change. What is the relationship of those to the knowledge generated by researchers?

If we consider the people involved, it is clear that it is only practitioners who have it in their power to change, and therefore to reform, practice. Sadly the researcher/practitioner relationship has frequently not been a close one, nor has the knowledge each generates. As Kilpatrick (1992) says: 'The actions of a practitioner who interprets classroom events within their contexts could not be further removed from the inferences made by a researcher caught up in controlling variation, quantifying effects, and using statistical models' (p. 31).

Can these two sets of activities be reconciled? More important perhaps, is whether practitioners and researchers are dealing with such different kinds of knowledge that communication becomes impossible.

Furthermore, the Research–Development–Dissemination model still lives in many assumptions about the relationship between research and practice, and reflects an assumed power structure which accords the researcher's agenda and actions greater authority than the practitioner's. The increasing moves to involve teachers in research teams are to be applauded, but currently only serve to reinforce the existing power structure. We hear little about researchers being invited to join teaching or curriculum planning teams. If the two knowledge domains are at present so mutually exclusive, then what hope is there for research to be influential in reforming practice?

Researchers clearly need to take far more seriously than they have done the fact that reforming practice lies in the practitioners' domain of knowledge. One consequence is that researchers need to engage more with practitioners' knowledge, perspectives, work and activity situation, with actual materials and actual constraints, and within actual social and institutional contexts. We will look at a good example of this later in the chapter.

There are some encouraging signs that researchers are engaging more with actual classroom events within actual classrooms. We are learning more about the teachers' knowledge, perspectives and so forth. The research agendas though are still dominated by the researchers' questions and orientations, not the practitioners'. Researchers tend to be interested in communication patterns, constructivist issues, group processes and so forth, any of which *may* produce ideas which could have reform implications. But we know precious little about *teachers*' perspectives on reform-driven issues which researchers could seriously address.

Dan Lortie's (1975) conclusions still ring true: 'Teachers have an in-built resistance to change because they believe that their work environment has never permitted them to show what they can really do' (p. 235). Lortie's view is that as a result, teachers often see the proposals for change made by others as 'frivolous' when they do not actually affect their working constraints.

Schools are more than just a collection of one-teacher classrooms. Consideration of the social dimension of mathematics education forces us to realize the institutional constraints which shape students' learning. Arfwedson (1976) analyzed the 'goals' and the 'rules' of a school, from the teacher's perspective, and showed that the goals of a school are not seen to be accompanied by sanctions since they are related to the pedagogical methods and attitudes adopted, whereas rules *do* carry sanctions (e.g., a teacher cannot disregard keeping to a timetable nor recording pupils' attendance with impunity). He also points out that the power of the teacher is somewhat superficial since there would appear to be inevitable conflict between rules and goals, and although the teacher has apparent pedagogical freedom, the rules impinge strongly. 'On the one hand the teacher is a part of the hierarchical power-structure of the school organization, on the other hand it is his duty to realize goals that are mainly democratic and anti-authoritarian' (pp. 141–142). Thus any desire on the part of the teacher to bring about change can become inhibited or destroyed.

Much clearly depends on the other practitioners who shape mathematics education. At the government, state, and local levels, there are curriculum and assessment designers. Textbook writers who receive governmental contracts are equally influential. School administrators, who may be part-time teachers, shape whole-school curricula and structure timetables and schedules which constrain what classroom teachers are able to do. Add parents, local employers, and politicians into the mix with their different agendas, and Arfwedson's and Lortie's comments assume an even greater significance.

Where is the research on the influence of different timetable and scheduling patterns on mathematical learning? Where are the researchers who are prepared to engage in the real practitioners' world of time constraints, local politics, and petty bureaucracies? There seems to be a certain amount of coyness on the part of many researchers, perhaps rationalized in terms of 'academic freedom', to join the practitioners' world. There is also a large amount of academic snobbery together with plenty of wishful thinking.

If researchers are to stand any chance of helping to reform practice, then they surely must enter the practitioners' world, derive more of their agendas from the problems of that world, conform more to practitioners' criteria and norms for solving those problems, and communicate more within practitioners' communication mechanisms, such as teachers' journals and newspapers.

Research Approaches and Practice

Are certain kinds of research activity better suited to improving practice than others? In Bishop (1992) I elaborated my ideas about the three components of the research process:

- Enquiry, which concerns the reason for the research activity. It represents the systematic quest for knowledge, the search for understanding, and gives the dynamism to the activity. Research must be *intentional* enquiry.
- Evidence, which is necessary in order to keep the research related to the reality of the mathematical education situation under study, be it classrooms, syllabuses, textbooks, or historical documents. Evidence samples the reality on which the theorizing is focused.
- Theory, which recognizes the existence of values, assumptions, and generalized relationships. It is the way in which we represent the knowledge and understanding that comes from any particular research study. Theory is the essential product of the research activity, and theorizing is, therefore, its essential goal. (p. 711)

Of the three main traditions influencing research (see Bishop 1992), the pedagogue and the scholastic philosopher seem to differ most markedly in respect of the relationship with practice (see Table 13.1):

Table 13.1

Theory	Goal of enquiry	Role of evidence	Role of theory
Pedagogue tradition	Direct improvement of teaching	Providing selective and exemplary children's behavior	Accumulated and shareable wisdom of expert teachers
Empirical scientist tradition	Explanation of educational reality	Objective data, offering facts to be explained	Explanatory, tested against the data
Scholastic philosopher tradition	Establishment of rigorously argued theoretical position	Assumed to be known. Otherwise remains to be developed	Idealized situation to which educational reality should aim

Source: Bishop 1992, p. 713.

The pedagogue tradition is overtly concerned with improving practice, while the scholastic philosopher is not. The pedagogue tends to involve teachers *in* the research process, whereas the scholastic philosopher tends to treat teachers as agents of practice who are largely irrelevant to the research process. The pedagogue tradition puts a large premium on 'knowing' the educational reality – the scholastic philosopher tradition takes that reality as a generalized assumption on which to base theorized possibilities.

Perhaps the pressure for increased effectiveness is reflected in the increasing dominance of pedagogue-influenced research approaches. As one example, we now find much less reliance on surveys and questionnaires and much more emphasis on case studies and anthropologically stimulated research.

Perhaps also the pressure is reflected in a tendency to choose research methods appropriate for a particular problem rather than to stay wedded to a particular research method. Begle's (1969) rallying call, to which many responded, is summed up in this statement:

> I see little hope for any further substantial improvements in mathematics education until we turn mathematics education into an experimental science, until we abandon our reliance on philosophical discussion based on dubious assumptions and instead follow a carefully correlated pattern of observation and speculation, the pattern so successfully employed by the physical and natural scientists. (p. 242)

As various writers in the *Handbook of Research on Mathematics Teaching and Learning* indicate, that view no longer predominates, largely because of the failure of the experimental method to produce the benefits sought by the providers of research funds, and by the educational system at large.

The question to reflect on is whether *any* one research method can meet the demands of improving the effectiveness of mathematics teaching. Recently, action research has been promoted as the only research approach which will have lasting effects in schools (Carr & Kemmis 1986). Action research, as described by critical-theorist proponents such as Carr, Kemmis and McTaggart, is concerned with research by people on their own work, with the explicit aim of improvement, and following an essentially critical approach to schooling: 'Philosophers have only interpreted the world in various ways ... the point is to change it' (Carr & Kemmis 1986, p. 156).

More than any other approach, action research takes 'change' as its focus and encourages practitioners of different kinds to research collaboratively their shared problems. As Kemmis and McTaggart (1988) point out, 'the approach is only action research when it is collaborative' (p. 5). Action research thus emphasizes group collaboration by the participants rather than the specific adoption of any one particular method. There is also a strong ideological component to action research, and it is thus more appropriate to refer to it as a 'methodology' rather than as one specific method (such as case study).

The action research methodology goes a stage further towards combining research and practice than other approaches where teachers join research teams. The latter practice, although encouraged by many, still tends to perpetuate the center–periphery model of educational change, defined by Popkewitz (1988) as consisting of the following stages:

1. initial research identifies, conceptualises and tests ideas without any direct concern for practice;
2. development moves the research findings into problems of engineering and 'packaging' of a program that would be suitable for school use;
3. this is followed by dissemination (diffusion) to tell, show and train people about the uses and possibilities of the program;
4. the final state is adoption/installation. The change becomes an integral and accepted part of the school system. (p. 131)

This is a model which mathematics education research has often adopted, either deliberately or accidentally. Indeed, some projects have stayed at the first stage, while others have moved to Stages 2 and 3, often in the form of textbooks or computer programs. Few, however, have reached Stage 4.

Partly the reasons are that the research has tended to focus mainly on the *learning* of the particular topic in question, or on the *curriculum* questions. If this happens, then there is no particular reason why any *teaching* approaches derived from or implied by the research will be at all successful.

Choosing to focus research on school-based curriculum topics does not necessarily produce improvements in teaching either. The content-oriented chapters in Grouws (1992) bear testimony to that:

1. After an exhaustive analysis of both the semantics of, and children's understanding of, rational number concepts, Behr et al. (1992) state that 'little is known about instructional situations that might facilitate children's ability to partition' (p. 316).
2. Concluding the chapter on algebra, Kieran (1992) says:

   > As we have seen, the amount of research that has been carried out with algebra teachers is minimal.... Teachers who would like to consider changing their structural teaching approaches and not to deliver the material as it is currently developed and sequenced in most textbooks are obliged to look elsewhere for guidance. (p. 413)

3. Concluding the chapter on geometry, Clements and Battista (1992) say: 'We know a substantial amount about students' learning of geometric concepts. We need teaching/learning research that leads students to construct robust concepts' (p. 457).
4. In the concluding section of the chapter on probability, Shaughnessy (1992) says: 'It is crucial that researchers involve teachers in future research projects, because teachers are the ultimate key to statistical literacy in our students' (p. 489).
5. Concluding the chapter on problem-solving, Schoenfeld (1992) says: 'There is a host of unsolved and largely unaddressed questions dealing with instruction and assessment' (p. 365). As Brophy (1986) says: 'Mathematics educators need to think more about *instruction*, not just curriculum and learning' (p. 325).

As was stated above, theory is the way in which we report the knowledge and understanding that comes from any particular research study. However, in relation to the theme of this chapter, the central issue about theory is to what extent should it shape the research itself? This issue highlights the role of theory in determining the research questions, in shaping the research process, and in determining the research method.

The thrust of the arguments so far in this chapter is that it should be the practitioners' problems and questions which should shape the research, not theory. So should research be any more than just practical problem-solving? And to what extent should the knowledge which is research's outcome be any more valid as a form of knowledge-for-change as any other form of knowledge?

Action research has certainly been interpreted by some as just practical problem-solving. However, as Ellerton et al. (1989) point out: 'It involves more than problem solving in that it is as much concerned with problem posing as it is with problem solving' (p. 285). They go on to show that action researchers do *not*

> identify with what has come to be known as 'the problem-solving approach' to achieving educational change. By this latter approach, a school staff identified problems which need to be addressed in their immediate setting: they design solutions to these problems, and in so doing, become trained in procedures for solving future problems. (p. 286)

For Carr and Kemmis (1986, p. 159), the term *research* implies that the perspectives of the participants in the research are changed. Thus, even in action research, the research process should be a significant learning experience for the participants. In that sense, the research problems and questions need to be couched in terms within the participants' schemes of knowledge. Theory enters through the participants' knowledge schemes, and insofar as these schemes involve connections with published theory, so will that theory play a part in shaping the research. It is this theory which offers the dimensions of generality which make the difference between research and problem-solving.

This pattern is well illustrated by a powerful study of mathematics teaching in first schools (Desforges & Cockburn 1987), which followed the analysis of Doyle (1986). In that sense, it built on a theoretical notion coming from earlier work, but the research questions are centered in the practitioners' world, particularly:

'What factors do teachers take into consideration in adopting management and teaching techniques and what factors force her to amend her goals or behaviors in the day-to-day world of the classroom?' (p. 23). The study describes and interprets the teachers' behaviors in the context of the teachers' reality and looks generally at the activities relating to stimulating higher-order thinking in mathematics, e.g. in problem-solving.

After documenting the teachers' practices in detail, the study concludes that it was clearly dealing with teachers of quality, who yet failed to deliver many of the aspirations which they and other mathematics educators fully endorsed. The authors ask:

> Why was there so much pencil and paper work and so little meaningful investigation? Why was there so little teacher-pupil and pupil-pupil discussion? Why was there so little diagnostic work? Why was the curriculum so dominated by formal mathematics schemes and so little influenced by children's spontaneous interests? Why did teachers with such an elaborate view of children's thinking cast their pupils into passive-receptive roles as learners or permit them to adopt such roles? (p. 125)

After further analysis the researchers point to the classroom and institutional realities which shape the practices, and comment that those realities are not designed for the conscious development of higher-order thinking. Indeed rather than criticizing the teachers for failing, they point out just how daunting it is to establish and sustain higher-order skills in a mathematics curriculum. The teachers' achievements are thereby that much more impressive. They say: 'We conclude that classrooms as presently conceived and resourced are simply not good places in which to expect the development of the sorts of higher-order skills currently desired from a mathematics curriculum' (p. 139).

I had reached the same conclusions in Bishop (1980) when I said:

> The problem is that classrooms appear not to be particularly appropriate environments in which to learn mathematics. Classroom learning can be characterized by the following constraints:
>
> (a) It must take place in a limited time
> (b) It is often incomplete learning
> (c) There are multiple objectives
> (d) The conditions are 'noisy'
> (e) The atmosphere is one of mutual evaluation
> (f) Presentation sequences are a compromise
> (g) Teaching is a stressful occupation
>
> Research is developing rapidly and our knowledge of learning is becoming more and more sophisticated. Meanwhile classrooms are becoming more of a challenge for teachers and many feel that the quality of teaching is declining. (pp. 339–340)

Desforges and Cockburn's (1987) theoretical analysis led them to conclude that, for practice,

> our prescriptions for change are directed more at those who provide – materially and especially conceptually – for practice and only tangentially at those who execute it. That is

> to say that in so far as contemporary mathematics teaching practices in the infant school may be seen to fall short of expectations, the burden of responsibility, in the terms of our analysis, lies with the educational managers who – whether deliberately or by default – provide the crucial psychological parameters of the teaching environment to which teachers and children alike must adapt. It is on these same groups that the onus for change must lie. (p. 143)

The Desforges and Cockburn study demonstrates vividly what researchers can contribute to the development of practice not only by contextualizing the research in the classroom realities, but also by couching the whole study in terms of practitioners' knowledge schemes. It also demonstrates that theory development is a goal, in that the study is both an analysis of practice and a search for explanations.

Perhaps one of the most important consequences for theory development is that researchers should pay more attention to synthesizing results and theories from different studies. As was said at the start of this chapter, the researchers at the ICMI Study Conference illustrated the tendency of researchers everywhere to analyze, critique and seek alternatives to each others' ideas, rather than trying to synthesize, build consensus, or recognize agreements. It is no good expecting teachers, or any other practitioners, to do the synthesizing, as they are frequently not the ones with access to the different ideas, results or approaches. An implication of a study like the one above is that it is the practitioners' epistemologies which should provide the construct base of the synthesized theory.

Researchers' Roles and the Practitioners' World

Research is big business in some countries, while in others it is another arm of government. In some situations, researchers can do whatever they like, while in others their practices are heavily proscribed either by external agencies or by ethical codes. Most would probably still yearn longingly for the academic ideal of the disinterested researcher, defining their own research in their relentless pursuit of knowledge. No researchers worth their salt would have any difficulty generating a research agenda for well into the 21st century. 'Knowledge for knowledge's sake' may sound old-fashioned but would still find appeal with many today.

However, the climate is changing. The days of the disinterested researcher are both ideologically and realistically numbered. For the big-ticket researcher, large research budgets are probably a thing of the past. In some countries, researchers have been deliberately marginalized from mainstream educational debate, while in others they have deliberately excluded themselves from it, and have found their research aspirations severely blunted. Increasingly, educational research and researchers are having to justify their continued existence in a world that is financially competitive, often politically antagonistic to institutionalized critique, and increasingly impatient with 'time-wasting' reflection and questioning.

If research is to have the kind of impact on practice and on the practitioners' world that many want it to have, then there is a clear need for much more disciplined enquiry into the practitioners' situation and an urgency to grapple with the issue of the theory/practice relationship. As a conclusion to this chapter and as a contribution to the debate, I offer the following ideas:

- *Researchers need to focus more attention on practitioners' everyday situations and perspectives.* The research site should be the practitioners' work situation, and the language, epistemologies, and theories of practitioners should help to shape the research questions, goals and approaches.
- *Team research by researchers/practitioners should be emphasized.* The work and time balance of the research activity will need to be negotiated, and the roles of the members clarified. The team should also include practitioners from other parts of the institution other than those whose activities are the focus. They are often the people who set the constraints on the development of teaching, as the study by Desforges and Cockburn (1987) showed. It seems to be of little value to involve them only at a dissemination stage, since their activities might well have contributed in an indirect way to the outcomes of the research.
- *The institutional context and constraints should be given greater prominence in research.* This is the 'practice' counterpart to the 'practitioner' point above. Institutions develop their own rules, history, dynamics and politics, and these need to be recognized and taken account of in the research.
- *Exceptional situations should be recognized as such, and not treated as 'normal' or generalizable.* Indeed, it is better to assume that every situation is exceptional, rather than assume it is typical. Typicality needs to be established before its outcomes can be generalized.
- *The process of educational change needs to be a greater focus in research in mathematics teaching.* It is rather surprising that, although many researchers assume a goal of change in their research, there has been relatively little research focus on the process of change itself.
- *Social and anthropological approaches to research should increase in prominence.* These approaches seem likely to offer the best way forward if researchers hope to make significant advances in how practitioners change their ideas and activities. Again it is no accident that they are already coming into greater prominence.
- *Conclusions and outcomes should be published in forms which are accessible to the maximum number of practitioners.* Researchers should resist the pressure to publish only in research journals, as these are rarely read by practitioners. If a team approach is adopted more frequently, then the practitioner members of the team can, and should, help with appropriate publication and dissemination.

It seems appropriate to finish this chapter by quoting a few more sentences from Desforges and Cockburn's (1987) study because they address the need for researchers to enter the practitioners' world, to admit their ignorance and to struggle to develop new theoretical interpretations:

> Rather than creating the aura that the only factor preventing the attainment of our aspirations in early years' mathematics teaching is the conservative practice of teachers, experts in the field should admit that they have yet to equip themselves – let alone the profession – with the conceptual tools adequate to the job. ... Such an admission might draw more first school

teachers into the kind of research work necessary. We have shown that teachers have a vast knowledge of children's responses to tasks. They are also very self-critical. Because they care about children it is very easy to make them feel guilty and feeling guilty they withdraw in the face of self-confessed experts. In this way researchers throw away their best resource, leave teachers open to cheap political jibes and make teaching more difficult. (p. 154)

References

Arfwedson, G.: 1976. 'Ideals and Reality of Schooling', in M. Otte, R. Bromme, D. Kallòs, U. Lundgren, T. Mies & D. Walker (eds.), *Relating Theory to Practice in Educational Research* (Materialien und Studien, Band 6). Institut für Didaktik der Mathematik, Universität Bielefeld, Bielefeld, Germany, 139–146.

Begle, E. G.: 1969, 'The Role of Research in the Improvement of Mathematics Education', *Educational Studies in Mathematics* **2** (2/3), 232–244.

Behr, M. J., Harel, G., Post, T. & Lesh, R.: 1992, 'Rational Number, Ratio, and Proportion', in D. A. Grouws (ed.), *Handbook of Research on Mathematics Teaching and Learning*, Macmillan, New York, 296–333.

Bishop A. J.: 1980, 'Classroom Conditions for Learning Mathematics', in R. Karplus (ed.), *Proceedings of the Fourth International Conference for the Psychology of Mathematics Education*, University of California, Berkeley, CA, 338–344.

Bishop, A. J.: 1992, 'International Perspectives on Research in Mathematics Education', in D. A. Grouws (ed.), *Handbook of Research on Mathematics Teaching and Learning*, Macmillan, New York, 710–723.

Brophy, J.: 1986, 'Teaching and Learning Mathematics: Where Research Should be Going', *Journal for Research in Mathematics Education,* **17** (5), 323–336.

Carr, W., & Kemmis S.: 1986, *Becoming Critical: Education, Knowledge and Action Research*, Deakin University Press Geelong, Victoria.

Clements, D. H., & Battista M. T.: 1992, 'Geometry and Spatial Reasoning' in D. A. Grouws (ed.), *Handbook of Research on Mathematics Teaching and Learning*, Macmillan, New York, 420–464.

Crosswhite, F. J.: 1987, 'Cognitive Science and Mathematics Education: A Mathematics Educator's Perspective', in A. H. Schoenfeld (ed.), *Cognitive Science and Mathematics Education.* Erlbaum, Hillsdale, NJ, 165–277.

Damerow, P., Dunkley, M. E., Nebres, B. F. & Werry, B.: 1986, *Mathematics for All*, Unesco, Paris.

Davis, R. B.: 1992, 'Reflections on Where Mathematics Education Now Stands and on Where it May Be Going', in D. A. Grouws (ed.), *Handbook of Research on Mathematics Teaching and Learning*, Macmillan, New York, 724–734.

Desforges, C. & Cockburn, A: 1987, *Understanding the Mathematics Teacher: A Study of Practice in First Schools*, Falmer, London.

Doyle, W.: 1986, 'Academic Work', *Review of Educational Research,* **53**, 159–200.

Ellerton, N. F.: 1986, 'Children's Made-up Mathematics Problems – A New Perspective on Talented Mathematicians', *Educational Studies in Mathematics* **17** (3), 261–271.

Ellerton, N. F., Clements, M. A. & Skehan, S.: 1989, 'Action Research and the Ownership of Change: A Case Study', in N. F. Ellerton & M. A. Clements (eds.), *School Mathematics: The Challenge to Change*, Deakin University Press. Geelong, Victoria, 184–302.

Freudenthal, H.: 1983, 'Major Problems of Mathematics Education', in M. Zweng, T. Green, J. Kilpatrick, H. Pollak, & M. Suydam (eds.), *Proceedings of the Fourth International Congress on Mathematical Education*. Birkhäuser, Boston, 1–7.

Grouws, D. A. (ed.): 1992, *Handbook of Research on Mathematics Teaching and Learning*, Macmillan New York.

Grouws, D. A., Cooney T. J., & Jones, D.: 1989, *Perspectives on Research on Effective Mathematics Teaching*, Erlbaum, Hillsdale, NJ.

Kemmis, S. & McTaggart, T. (eds.): 1988, *The Action Research Planner*, Deakin University Press, Geelong, Victoria.

Kieran, C.: 1992, 'The Learning and Teaching of School Algebra', in D. A. Grouws (ed.), *Handbook of Research on Mathematics Teaching and Learning*, Macmillan, New York, 390–419.

Kilpatrick, J.: 1981, 'The Reasonable Ineffectiveness of Research in Mathematics Education', *For the Learning of Mathematics,* **2** (2), 22–29.

Kilpatrick, J.: 1992, 'A History of Research in Mathematics Education', In D. A. Grouws (ed.), *Handbook of Research on Mathematics Teaching and Learning*, Macmillan, New York, 3–38.

Lortie, D. C.: 1975, *Schoolteacher: A Sociological Study*, University of Chicago Press; Chicago.

Popkewitz, T. S.: 1988, 'Institutional Issues in the Study of School Mathematics', *Educational Studies in Mathematics* **19** (2), 221–249.

Research Advisory Committee: 1990, 'Mathematics Education Reform and Mathematics Education Research: Opportunities, Obstacles and obligations', *Journal for Research in Mathematics Education* **21** (4), 287–292.

Research Advisory Committee: 1993, 'Partnership', *Journal for Research in Mathematics Education* **24** (4), 324–328.

Schoenfeld, A. H.: 1992, 'Learning to Think Mathematically: Problem Solving, Metacognition, and Sense Making in Mathematics', in D. A. Grouws (ed.), *Handbook of Research on Mathematics Teaching and Learning*, Macmillan, New York, 334–370.

Shaughnessy, J. M.: 1992, 'Research in Probability and Statistics: Reflections and Directions', in D. A. Grouws (ed.), *Handbook of Research on Mathematics Teaching and Learning*, Macmillan, New York, 334–370.

Alan J. Bishop
Faculty of Education,
Monash University,
Clayton,
Victoria 3168,
Australia

Chapter 14
Practicing Research and Researching Practice

Jeremy Kilpatrick

In 1992, the Executive Committee of the International Commission on Mathematical Instruction (ICMI) appointed a program committee to organize and conduct an ICMI Study on the topic "*What is research in mathematics education, and what are its results?*" The committee was charged with preparing a discussion document that would frame the topic, conduct a study conference at which the topic would be discussed and analyzed, and produce a report to be published by Kluwer Academic Publishers in the New ICMI Study Series. The charge included the request that major outcomes of the study be presented at the 1994 International Congress of Mathematicians (ICM) in Zürich.

The discussion document (Balacheff et al., 1998) was published in several journals for mathematics educators in late 1992 and early 1993. It presented a framework for discussion in the form of five questions:

1. What is the specific object of study in mathematics education?
2. What are the aims of research in mathematics education?
3. What are the specific research questions or *problématiques* of research in mathematics education?
4. What are the results of research in mathematics education?
5. What criteria should be used to evaluate the results of research in mathematics education?

The document also contained a call for papers, and those papers and other expressions of interest were used to organize a study conference that was held from 8 to 11 May 1994, primarily on the University of Maryland campus in College Park but with a half-day symposium at the National Academy of Sciences building in Washington, DC. The conference, with 81 invited participants, comprised plenary sessions, working groups on each of the five questions above, and paper sessions for discussion of examples of research relating to the questions.

J. Kilpatrick
105 Aderhold Hall, University of Georgia, Athens, GA 30602-7124, United States of America
e-mail: jkilpat@uga.edu

P. Clarkson, N. Presmeg (eds.), *Critical Issues in Mathematics Education*,

Summarizing the ICMI Study Conference

From almost the beginning of the planning for the conference, the program committee had proposed that Alan Bishop be invited to give a plenary talk at the final session of the conference, after the working groups had presented their reports. Alan was asked, and agreed, to summarize what he had seen and heard over the 4 days. At that session, which he entitled "Disagreements, dichotomies, and developed ignorance," he spoke for 40 minutes using nine overhead transparencies. In his summary, he: (a) questioned whether we had assembled the essential ingredients for a research conference in mathematics education; (b) characterized the emphases and silences he had noted in the sessions; (c) identified metaphors that conferees had used or that came to mind; (d) listed dichotomies he had noticed (Fig. 14.1), while simultaneously asking whether they might be false; (e) deconstructed *mathematics education,*

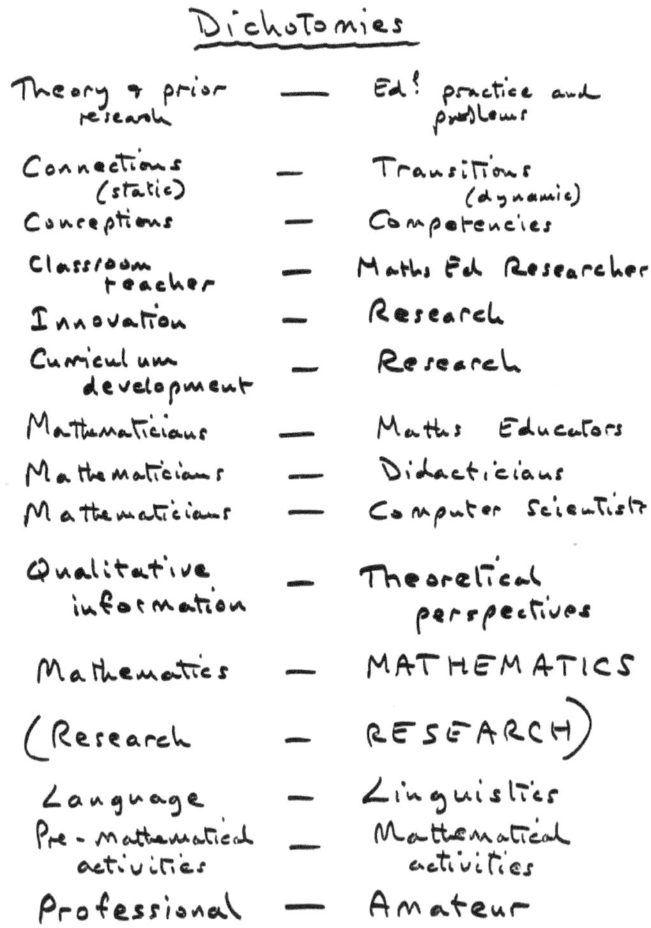

Fig. 14.1 An overhead transparency from Alan Bishop's 1994 talk

mathematics, and *education*; (f) raised issues about the relations among problems, research methods, and theories and among teachers, researchers, and the educational system; (g) asked what it would be reasonable for the study to produce and for the ICMI to do with respect to research; and (h) noted that we needed to collaborate rather than compete as we search for meaning and that we needed to transcend "our developed ignorance" as researchers in mathematics education.

The conferees responded enthusiastically to Alan's wide-ranging, perceptive summary, which nicely pulled together the many diverse strands of their deliberations. Asked to provide a copy of Alan's transparencies, I had them typed up and sent out with a cover letter thanking the conferees for their participation and encouraging them to submit new or revised papers reflecting thoughts stimulated by the conference.

Pressure for Effectiveness

Before the conference, readers of the discussion document who were interested in contributing to the ICMI study had been invited to submit papers on specific problems or issues raised by the document. Bishop and Dudley Blane had submitted a paper entitled "Responding to the Pressure for 'Effectiveness' in Mathematics Teaching," which was included in the 430-page collection of background papers given to each conferee (Bishop & Blane, 1994). The central argument of the paper was that researchers needed to address more seriously the demand for knowledge about the effectiveness of mathematics teaching. Only if researchers conducted more disciplined inquiry into the practitioners' world and dealt more skillfully with the nexus between theory and practice would research contribute to any reform of that practice. In the end, several papers subsequently published in the study report (Sierpinska & Kilpatrick, 1998) addressed aspects of the same theme, including several of the working group reports and the chapters by Paolo Boero and Julianna R. Szendrei, Gérard Vergnaud, Ferdinando Arzarello and Mariolina Bartolini-Bussi, Koeno Gravemeijer, Gillian Hatch and Christine Shiu, and Claire Margolinas, among others.

Bishop's Chapter

When the editors of the ICMI study report (Anna Sierpinska and I) began to assemble the manuscripts for publication, we realized that a different organization would be needed than had been planned. We had thought that we could use the five study questions to organize the chapters, but that turned out to be impossible. Although some chapters focused on a single question, many others addressed all five, and some dealt with other issues. We organized the book instead into six parts:

I. The ICMI Study Conference
II. Mathematics Education as a Research Discipline

III. Goals, Orientations and Results of Research in Mathematics Education
IV. Different Research Paradigms in Mathematics Education
V. Evaluation of Research in Mathematics Education
VI. Mathematics Education and Mathematics

We had expected that Bishop's chapter would conclude the report, but it did not fit well at the end. Instead of being a summary, it included only a short section (just over a page) at the beginning that dealt with the conference. The remainder was a revised version of the paper by Bishop & Blane (1994). We contemplated putting the chapter in Part III of the report but eventually decided it worked better at the end of Part I.

It is important in reading the chapter, therefore, to understand that the main message was one that Bishop (1998) brought to the conference and not one that he discovered there. His tone is that of a stern parent reproving researchers in mathematics education for their insularity and ineffectiveness. He addresses two sorts of effectiveness: the demands by the public in many countries that mathematics teaching be more effective and the need for research in mathematics education to be more effective in reforming practice. Bishop makes several paradoxical observations in the chapter. One is that although research has contributed to the press for effective teaching by providing evidence on competence, raising expectations about all students' potential for achievement, and helping to generate alternative teaching approaches (p. 34), research appears to have had little influence thus far on changing teaching practice. Research in mathematics education is apparently both potent and impotent. Another seeming paradox concerns the role of theory. On the one hand, Bishop argues, "it should be the practitioners' problems and questions which should shape the research, not theory" (p. 40). On the other hand, "theory enters through the participants' knowledge schemes, and insofar as these schemes involve connections with published theory, so will that theory play a part in shaping the research" (p. 41). Theory should both not shape and shape research.

Bishop ends the chapter by citing a study by Desforges and Cockburn (1987) that "demonstrates theory development is a goal, in that the study is both an analysis of practice and a search for explanations. [Moreover] it is the practitioners' epistemologies which should provide the construct base of the synthesized theory" (p. 42). Whatever role theory plays in their work, Bishop (1998) concludes, researchers need to enter the practitioners' world.

Teachers and Research

The Royaumont Meeting

The theory-practice nexus has preoccupied Alan throughout his career, and he has long believed that teachers should play a much greater role in research in mathematics education than they typically have. I recall learning of that preoccupation and belief when Alan and I first met, in September 1971. We had been invited to Royaumont Abbey, outside Paris, where we spent 2 weeks with a 16-member international group of mathematics educators preparing a document later published as the third volume of Unesco's *New Trends in Mathematics Teaching* (Fehr & Glaymann, 1972). I hastily point out that this meeting should not be confused with the Organisation for European Economic Co-operation seminar at Royaumont in 1959 at which Jean Dieudonné uttered his famous cry, "*À bas Euclide!*" Our work was quieter and less fiery.

Alan and I were two rookies amid a crowd of eminent heavy hitters such as Hans Freudenthal, Maurice Glaymann, Anna Sophia Krygowska, Georges Papy, André Revuz, Willy Servais, and Hans-Georg Steiner. The two of us had apparently been invited because of some work we had each done in evaluation and in research, and we ended up working together on the two chapters that dealt with those topics.

In the following quotations from the chapter entitled "Research in Mathematical Education" (Fehr & Glaymann, 1972), even though the chapter is unsigned, one can hear Alan's voice delineating the role of the practitioner in research:

> One trend that has emerged in some countries in the last few years is a recognition of the validity of [the] "teacher's view" of research in mathematical education. Research does seem to be moving into the classroom more and more. Some researchers at least have begun to see that there are many avenues of research, but that all educational research must at some point make contact with the concerns of teachers or it will come to nothing. (p. 129)
>
> In all countries there is an untapped resource of teachers who have scholarly inclinations, who are interested in research, and who—perhaps given some assistance—could conduct useful research studies in their classrooms. A challenge to researchers in mathematics education in the next decade or so will be to find ways to make their own research more relevant, and the work of teachers more productive, by enlisting teachers as partners in the research enterprise. (p. 136)

Additional indications of Alan's concern with practitioners come from a paper (Bishop, 1972) summarizing research trends in Europe that he prepared for the Royaumont meeting:

> In much of the traditional research on teaching, the individual teacher has tended to be ignored and any work which can shed light on teachers' values and decisions as they affect the minute-by-minute interactions in the classroom is to be welcomed. (p. 16)
>
> To summarize then, it is clear that the research in Europe is less concerned at the present time with developing a theory or a 'science' of mathematical education than with pursuing those areas which are closely related to the classroom, namely the curriculum, the child and the teacher. . . . [The most important reason for this pursuit] could be the feeling that education begins, and ends, in the classroom and that any researcher who loses touch with that 'real' situation is in danger of losing touch with education. (p. 17)

After Royaumont, I spent my 1973–1974 sabbatical year at Cambridge University, where I observed firsthand Alan's work with prospective and practicing teachers. I also got acquainted with his work on spatial abilities and on teachers' decision making. In his courses at the university as well as his supervision of novice teachers, he put into practice his belief that the teacher's role included not simply being familiar with research but also participating in it.

Papers From 1977

I have yellowing copies of two papers from 1977 that elaborate some of Bishop's perspectives on teachers and research. They come from his first trip to Australia and were delivered at conferences there. The first (Bishop, 1978) was delivered to a seminar for mathematics teachers in Melbourne, sponsored by the Mathematical Association of Victoria (MAV) in October 1977, and reprinted the following year in the MAV journal. Bishop identifies three trends in mathematics teaching: (a) making mathematics accessible to a wider school audience, (b) moving away from searching for a best method and toward respecting the teacher's individuality, and (c) introducing realistic, rather than pure, mathematics into the syllabus. In discussing research that supports the second trend, Bishop says,

> In those studies where teaching is the focus, it is the teacher who is given close-scrutiny, to try to uncover more about how people actually manage to teach. So we don't see so much emphasis on teaching-methods for example, we see instead research which looks at how teachers control and manage classrooms, how they make their decisions in class, how they characterize the pupils and what effect these characterizations have, how they communicate both verbally and non-verbally. (p. 9)

At the same time that he supports individuality in teaching, however, Bishop is quick to observe the need to "guard against innovation without evaluation" (p. 8). He calls for criteria for judging teaching that will suit the changing conditions of mathematics teaching.

The second paper (Bishop, 1977) was the opening plenary address to the annual conference of the MAV at Monash University in December 1977. The title is "On loosening the constructs" (although the published paper is given the mistaken title "On loosening the contents"), and the paper is an introduction to the work of George A. Kelly (1955), developer of the psychology of personal constructs, a theory of personality functioning that views learners as scientists who create constructs of their world and then test them for validity. If you want to understand a child's behavior, says Kelly, try to find out the question he or she is asking, the hypothesis he or she is testing. Bishop shows how Kelly's theory can be applied to mathematics teaching and learning.

One of Bishop's (1977) most arresting arguments is the following:

> I'm often asked about research and what results we should take note of. But I don't think the results are the important things. Too many people have that wrong view of educational research—the problem they say (especially if they're researchers) is not with the quality of the research itself but with the dissemination of the results. . . . No, in my view, there's more useful material than that in research. (p. 2)

He goes on to talk about the material that teachers can borrow from researchers: their procedures, data, and constructs—and not their results. And he points out the value of becoming aware of how your own personal constructs may be blinkering you, keeping you from soaking up new experiences:

> Theories and constructs are a bit like spectacles—some help you to see more clearly the object you are concerned with, while others merely give you a foggy, blurred image. Change

> the object of your concern, however, and the second pair of spectacles might be more useful. (p. 4)

I have long been inspired by this paper, and I borrowed shamelessly from it when I was asked to address the Canadian Mathematics Education Study Group (Kilpatrick, 1981). Here is part of what I said, some of which goes beyond Bishop's claims:

> The most important aspect of a research study is the construct and theories used to interpret the data. A landmark research study is one that confronts us with data analyzed and organized so as to shake our preconceptions and force us to consider new conceptions. . . . This view suggests why teachers should be active researchers, why they should develop a research attitude. Teachers should not stop at being borrowers; they should become collaborators. (p. 27)

Throughout Bishop's (1977) paper, he calls on teachers to engage in research while simultaneously encouraging them to keep matters in perspective.

Mathematicians and Research

A notable feature of Bishop's (1998) article for the ICMI study volume is that it nowhere mentions mathematicians. Moreover, the only overhead transparency used in his summary talk that contains the word *mathematicians* is the one shown in Fig. 14.1. The word is used there in contrast with mathematics educators, didacticians, and computer scientists—dichotomies between groups of people but not dichotomies that concern research. Neither at the conference nor in his paper did Bishop address the issue of what mathematicians might understand research in mathematics education to be or how they might make use of it.

In a way, that lacuna is curious. The ICMI study on research was prompted by the concerns of mathematicians who wanted to understand better the field of mathematics education. The problem was not simply that at the 1988 International Congress on Mathematical Education (ICME), "there was a general feeling that mathematics educators from different parts of the world, countries, or even areas of the same country often talk past one another" (Balacheff et al., 1998, p. 3). There had also been for some time "a need to explain the domain to representatives of other scientific communities, among which the community of mathematicians seems to be the most important" (p. 4). The mathematicians on the ICMI Executive Committee posed the central question, "What is research in mathematics education?" as well as the follow-up question, "And what are its results?" The final section of the ICMI study volume (Sierpinska & Kilpatrick, 1998) contains five articles on topics relating to mathematicians and research in mathematics education. It is odd that even though the ICMI study addressed other audiences as well, Bishop apparently did not see it as being aimed primarily at mathematicians.

Among the people who try to understand and use research in mathematics education are mathematicians in colleges and universities. They are practitioners in a different sense than schoolteachers are. Nonetheless, they are as likely as teachers are to think of results as the central characteristic of research. ICMs are filled with reports of recent results of mathematics research, so why are ICMEs not filled with reports of the results of research in mathematics education? Perhaps mathematicians need to hear, as much as teachers do, Bishop's (1977) argument that they should be taking from research not so much the results as the methods, data, or constructs.

Conclusion

It is no surprise that the first dichotomy on Alan's list in Fig. 14.1 is that between "theory and prior research" and "educational practice and problems." The nexus, complementarity, tension, and even opposition between theory and practice have been at the heart of Alan's concern with research in mathematics education. Throughout his long and productive career, he has continually reminded us that just as teachers are crucial to the teaching and learning of mathematics, so they are crucial to why, what, where, and how research is done on that teaching and learning.

References

Balacheff, N., Howson, A. G., Sfard, A., Steinbring, H., Kilpatrick, J., & Sierpinska, A. (1998). Discussion document. In A. Sierpinska, & J. Kilpatrick (Eds.), *Mathematics education as a research domain: A search for identity* (pp. 3–8). Dordrecht: Kluwer.

Bishop, A. J. (1972, Spring). Trends in research in mathematical education. *Mathematics Teaching, 58*, 14–17.

Bishop, A. (1977). On loosening the contents. In L. Murray (Ed.), *Meaningful mathematics* (pp. 1–4). Melbourne: Mathematical Association of Victoria.

Bishop, A. (1978). Trends in mathematics teaching. *Vinculum, 15*(1), 4–14.

Bishop, A. J. (1998). Research, effectiveness, and the practitioners' world. In A. Sierpinska, & J. Kilpatrick (Eds.), *Mathematics education as a research domain: A search for identity* (pp. 33–45). Dordrecht: Kluwer.

Bishop, A. J., & Blane, D. C. (1994, May). Responding to the pressure for "effectiveness" in mathematics teaching. In International Commission on Mathematics Instruction, *What is research in mathematics education, and what are its results?* (Background papers for an ICMI Study Conference), (pp. 52–68). University of Maryland, College Park.

Desforges, C., & Cockburn, A. (1987). *Understanding the mathematics teacher: A study of practice in first schools*. London: Falmer.

Fehr, H., & Glaymann, M. (Eds.) (1972). *New trends in mathematics teaching* (Vol. 3). Paris: Unesco.

Kelly, G. A. (1955). *The psychology of personal constructs*. New York: Norton.

Kilpatrick, J. (1981). The reasonable ineffectiveness of research in mathematics education. *For the Learning of Mathematics, 2*(2), 22–29.

Sierpinska, A., & Kilpatrick, J. (Eds.) (1998). *Mathematics education as a research domain: A search for identity*. Dordrecht: Kluwer.

Chapter 15
Reflexivity, Effectiveness, and the Interaction of Researcher and Practitioner Worlds

A Reflection on Bishop's "Research, Effectiveness, and the Practitioner's World" in the Light of a Quarter-Century of Systemic Improvement Effort in English Mathematics Education

Kenneth Ruthven

Introduction

Bishop's paper on "Research, effectiveness, and the practitioners' world" was published in 1998. It originated as a summary reaction to the ICMI study conference on "What is research in mathematics education, and what are its results?" held in 1994. As Bishop concedes, what he perceived as "researchers' difficulties of relating ideas from research with the practice of teaching and learning mathematics" (p. 189) may have been encouraged by the rather introverted character of this particular occasion. It is not surprising, after all, that a three-day meeting with a declared intention "to bring together representatives of . . . different groups of researchers, allow them to confront one another's views and approaches, and seek a better mutual understanding of what we might be talking about when we speak of research in mathematics education" (Sierpinska & Kilpatrick, 1998, p. 3) encouraged inward attention to intellectual differences rather than an outward focus on practical implications. Nevertheless, Bishop suggests that even when researchers do address issues of practice, the danger is that they do little more than provide evidence of problems and raise expectations about improvement, creating pressure for change rather than providing guidance for it. The central thrust of his paper is that researchers in mathematics education need to take (more) account of "the practical concerns of teachers" and to respond (better) to "pressures for more *effective* modes of mathematics education" (p. 190).

K. Ruthven
Faculty of Education, University of Cambridge, 184 Hills Road, Cambridge CB2 8PQ, United Kingdom
e-mail: kr18@hermes.cam.ac.uk

P. Clarkson, N. Presmeg (eds.), *Critical Issues in Mathematics Education*,

While Bishop's paper does allude to a political dimension to mathematics education, it largely passes over the powerful part that official agencies and policy-making bodies play in shaping educational reform, in supporting educational research, and in mediating between the two. Notably, government sponsorship of educational research usually reflects a desire for outcomes which can illuminate and advance the development of its policies and the changes in practice that these imply. Equally, while Bishop's paper acknowledges that "[m]uch clearly depends on the other practitioners who shape mathematics education" (p. 193), such as curriculum designers, textbook writers, external examiners, and school administrators, it gives little consideration to the role of intermediary bodies and organisations through which official policy is implemented and established practice adapted accordingly, and the part played in these processes by educational research. Moreover, where mathematics education is concerned, there is a considerable overlap in most educational systems between those who carry out teacher education and professional development on the one hand and educational research on the other; in particular, research in mathematics education would not survive in most universities without the financial base provided by these other activities. This paper will argue that, under such circumstances, official agencies and policy-making bodies exert as important an influence on the researchers' world as they do on the practitioners', shaping the kinds of educational research, developmental activity, and other forms of researcher engagement with practice which can be undertaken.

Indeed, the relative silence of Bishop's paper in this respect is surprising given Alan's personal contribution to leading one of the commissioned research reviews (Bishop & Nickson, 1983) that informed the work of the Cockcroft Committee in the early 1980s. So my argument will take Cockcroft as the starting point for a quarter century of endeavour to improve the quality and effectiveness of professional practice in English school mathematics in the light of relevant educational research (see also Ruthven, 1999), influenced by wider thinking about strategies for the development of "good practice" in teaching and learning (see also Ruthven, 2005). I will use the main lines of development over this period as a concrete counterpoint to assist reflection on some of the central ideas of Bishop's paper.

The Rise and Rebuff of Reflective Practice

The Rise

The Cockcroft Committee was a very significant attempt to use insights from educational research to analyse and address concerns about the quality and effectiveness of professional practice in English school mathematics. Bishop and Nickson's substantial review was one of three on different aspects of teaching and learning mathematics commissioned by the Department for Education on behalf of the Committee. In addition, the Department funded studies of the mathematical needs of employment and of everyday life, and a survey of relatively new entrants to mathematics teaching, to fill gaps in the available research identified by the Committee. The resulting Report (Cockcroft, 1982) sought to build professional consensus around a cogent system of ideas aimed directly at improving practice, explicitly informed by these thorough reviews of earlier developments and existing research. Equally, implementation of the Report's recommendations was marked by an unprecedented strengthening of intermediary capacity in two forms. First, there was support for substantial development projects, typically on an action-research model with strong teacher participation. Second, a national infrastructure was established to disseminate the Cockcroft ideas for reform through a network of advisory teachers, colloquially know as "mathematics missionaries". This precedent was surely in Bishop's mind when he proposed not only that "researchers should pay more attention to synthesizing results and theories from different studies" (p. 198), but that "it is the practitioners' epistemologies which should provide the construct base of the synthesized theory" (p. 198).

The Report identified a basis for "good practice": notably that mathematics teaching at all levels should include opportunities for exposition, discussion, practical work, problem solving, investigation, consolidation and practice (Cockcroft, 1982, para. 243). But, as the "missionary" metaphor signals, there was a tension within post-Cockcroft initiatives between actively promoting the innovative elements of this formulation and honouring the Report's reluctance "to indicate a definitive style for the teaching of mathematics" on the grounds that this was "[n]either desirable [n]or possible" (para. 242). Rather, the Report suggested that: "Approaches to the teaching of a particular piece of mathematics need to be related to the topic itself and to the abilities and experience of both teachers and pupils" (para. 242). Thus, some of the main post-Cockcroft initiatives involved networks of school-based teacher researchers working together under the leadership of university-based mathematics educators, influenced by ideas of action research and reflective practice. The action research approaches adopted in these initiatives corresponded closely with Bishop's view that "practitioners' problems and questions . . . should shape the research, not theory" (p. 198), so that "the research process should be a significant learning experience for the participants" (p. 196), and "theory enter[] through the participants' knowledge schemes" (p. 197).

A significant example was the Raising Achievement in Mathematics Project [RAMP], developed first in one region (LAMP, 1987), and then extended nationally (RAMP, 1991). While the project aimed to develop and encourage "good practice" along the lines recommended by the Cockcroft Report, it was based on the principle that "improvements and change can only be sustained if teachers in the classroom believe in and support the developments taking place" so that "dissemination must always be firmly rooted in the personal experiences of teachers in their classrooms" (LAMP, 1987, pp. 81–82). Within the project, then, "teacher-researchers explored possibilities and ideas within their own classrooms, involved their colleagues through discussion and collaborative teaching and kept personal records". In particular, through "individuals writing and talking about their own situations and experiences in a personal and uninhibited way", the project produced "case studies" which although "by their nature . . . not prescriptive" could suggest "ingredients which other teachers may identify as being transferable to their own classroom". The goal was that "through such discussion and personal experimentation the processes of questioning, experimenting, reflecting and evaluating become embedded in a teacher's practice", to form a "cycle [which] is far from being a 'clinical' or 'academic' one" because of the way in which it calls for "teachers to reflect so intensely on their own classroom practice, beliefs and assumptions" (LAMP, 1987, p. 7). In line with this appeal to personal experience and social influence, the RAMP project sought to spread its "varied and continuously growing accumulation of knowledge and experience" by encouraging participants to initiate similar developmental work with other teachers and pupils, so as to "form a network which grows as teachers make new contacts and form new working groups" (LAMP, 1987, p. 7). Indeed, the Foreword to the report of the first phase of the project illustrates its epistemic commitment to a social process of knowledge diffusion in which local validation through personal activity and judgement are viewed as crucial:

> The nature of the report reflects the quality of all this ongoing personal classroom-based research. The recommendations for action and the statements made are careful distillations of these experiences. The report is hence an invitation to those concerned with education at *all* levels to experiment, discuss, debate, strengthen and refute as a result of their *own* experiences. It is about taking action. (LAMP, 1987, p. vii)

This emphasis on teacher involvement in sustained developmental activities was characteristic of post-Cockcroft initiatives intended to realise reforms in curriculum, pedagogy and assessment along the lines proposed by the Report. Nevertheless, as wider stocktaking of practitioner action research has concluded (Elliott & Sarland, 1995; Elliott, Maclure & Sarland, 1997), while it can serve as an effective means of engaging teachers in professional development, the challenge of bringing participants to rigorously examine – in Bishop's terms – "their own work, with the explicit aim of improvement, and following an essentially critical approach to schooling" (p. 194) is a considerable one. While "teachers often see the proposals for change made by others as 'frivolous' when they do not actually affect their working constraints" (p. 192), the changes that schools and teachers are able to envisage and effect often reflect assumed constraints on, and associated patterns of, practice which have become professionally reified. Thus while Bishop suggests that "it is only practitioners who have it in their power to change . . . practice" (p. 191), he also highlights "the classroom and institutional realities which shape th[is] practice" (p. 197). Indeed, the shift in such 'realities' within the English system over the decade between publication of the Cockcroft Report in 1982 and Alan's departure for Australia in 1992 serves to emphasise not only their influence but their mutability.

The Rebuff

By the close of this period, the type of developmental activity receiving official support had changed markedly, as government grew sceptical of the devolved model of distributed professional leadership for educationally progressive reform associated with the post-Cockcroft initiatives in mathematics (and with wider educational innovation across the school system as a whole). Equally, the strongly developmental emphasis of these initiatives meant that there was little hard evidence and few cumulative insights with which to counter such scepticism:

> It is clear that these projects did impact on teachers and curricula, but, unfortunately, even where careful evaluations were carried out these were not reported in 'refereed journals', subject to peer review. In this sense, then, accumulated knowledge and experience is not available to be passed on and every new project 'reinvents the wheel'. Properly planned and funded evaluations should be a feature ... in the future so that succeeding curriculum developments can build on the strengths and address the weaknesses of previous innovations. (Askew & Wiliam, 1995, p. 198)

The 1987 Education Reform Act [ERA] laid the ground for a radical change in English schooling towards a centralised model based on unprecedented standardisation and regulation. Although the initial plans devised by the National Curriculum Mathematics Working Group envisaged a continuation of post-Cockcroft notions of "good practice", these plans were undermined as the government embraced an increasingly "back to basics" stance on educational matters, which included reverting to more reductive models of school mathematics (Dowling & Noss, 1990; Brown, 1993). Symptomatically, whereas the report on the first phase of RAMP (LAMP, 1987) was accorded some official standing through publication as a government document, the report on the second phase was not treated in the same way (RAMP, 1991). By that time, schools and teachers were being pressed to "deliver" a national curriculum and become "compliant" with national regulations, reinforced by high-stakes systems of school and teacher evaluation through regular pupil testing and school inspection.

The pattern of continuing professional development for teachers shifted markedly. The majority of provision was now managed directly by government agencies, and this focused on familiarising teachers with the new curriculum and assessment regime. The primary mechanism for professional development now became the distribution of official documents and guidance materials, supported by a "cascade" approach in which schools nominated subject leaders to attend organised training sessions with a view to their then leading similar activity with their school colleagues, sometimes with help from local authority advisory teachers. The role of universities in professional development was much reduced; the reflexivity which was the hallmark of their approach did not fit this emerging technocratic model for systemic improvement in education. Indeed, as its educational policy grew more directive, the government became, to echo Bishop's words, "politically antagonistic to institutionalized critique, and increasingly impatient with 'time wasting' reflection and questioning" (p. 198). Government politicians adopted a "discourse of derision", lambasting university-based teacher education and educational research, with one minister memorably characterising the results of some of the very work that his predecessors had commissioned as "barmy theory" and "elaborate nonsense".

In initial teacher education, then, new school-based routes were established and encouraged by government. However, severe teacher shortages and limited school capacity meant that the university contribution in this area could not be wholly discounted. Rather, government strengthened recent reforms which had introduced official regulation and evaluation of initial teacher education. In addition to lecturers being required to undertake "recent and relevant experience" as school teachers, student teachers following what was now restyled as "teacher training" were obliged to spend the great majority of their programme on school placement. Even more significantly, the collegial style which had persuaded universities to accept the introduction of official inspection of their courses of teacher education was set aside by government in favour of a much more authoritarian approach, and these programmes now became subject to detailed prescription and tight regulation in much the same way as schools. Alan encountered these reforms directly as a teacher educator on the postgraduate course in Cambridge that we taught together. Although widely perceived as acts of political spite, they did provide useful scope for small-scale, classroom-based research by lecturers (Bishop, 1991; Ruthven, 1989) and new opportunities to enhance the quality of teacher preparation on a "practical theorising" model (McIntyre, 1995; Ruthven, 2001).

The overall effect of these shifts in policy was to produce a marked realignment of the relationship between government, universities and schools, marginalising the more analytic and reflexive contribution of university-led educational research, teacher education and professional development to political and professional thinking. Whereas many university-based mathematics educators – most of whom brought to this work a successful professional history as schoolteachers and a strong commitment to educational improvement – had been deeply involved in the developmental activity and design research surrounding the Cockcroft review and the subsequent implementation of its recommendations, these new conditions of tight government regulation and evaluation made it difficult to carry out independent work of this type. Not surprisingly, many university-based mathematics educators adapted to this changed situation by pursuing lines of research and scholarship that allowed them to express professional detachment from the reforms. Perhaps much of the passion with which Alan wrote his paper arose from his experience as President of the Mathematical Association during this politically charged period, concerned with the capacity of the profession to engage with these reforms, and their impact on it.

The Rise and Redux of Effective Practice

The Rise

Thus, by the time that Alan left England in 1992, "increasing pressures for more *effective* modes of ... education" (p. 190) had come to dominate schooling policy, with external testing of pupils in the process of being introduced at four points between ages 7 and 16. Under a policy of "open enrolment", schools competed to attract pupils (with a school's funding related to its success in recruitment) in a "market" designed to be strongly influenced by the annual publication of "league tables" intended to inform prospective pupils and their parents about the relative success of schools on key "performance indicators" (notably the test and examination performance of their students). Increasingly, too, schools were competing to attract and retain teachers against a background of intensifying shortages in subjects such as mathematics. In trying to raise their measured "effectiveness", common strategies adopted by schools were to teach to the test; to give pupils more regular practice of test-taking; to give special attention to those pupils working close to the "pass/fail" boundaries for testing; and to prioritise classes containing such pupils when assigning teachers.

In 1997, a change of government brought a further intensification of these pressures through the setting of ambitious targets, increasing annually, for the proportions of students achieving benchmark levels of performance in each school. However, the new government was also determined to exercise a more direct influence on matters of classroom pedagogy and school management, encouraged by the confidence with which some educational figures argued that it was possible to identify effective teaching methods and implant them in schools. One of these – a leading researcher in school effectiveness and improvement – was appointed by the new government to chair its "Numeracy Task Force". Also amongst the members of the Task Force was a highly experienced professional leader in mathematics education, who was now directing what was already termed the "National Numeracy Project", established within the Department for Education by the outgoing government to develop, trial and refine "interactive whole-class" teaching methods at primary-school level. The work of the Task Force was closely "observed" by eight representatives from government agencies with a strong interest in its proposals, including four from the newly created "Standards and Effectiveness Unit" within the Department for Education, which would take responsibility for implementing the resulting National Numeracy Strategy, initially in primary schools, but rapidly extended to lower secondary.

The business of the Task Force was highly politicised. Not only was it charged with quickly producing a viable strategy capable of rapid implementation, but it was constrained in the options it could seriously consider by the ideological unacceptability of many of them to government ministers and their officials, regardless of what research might indicate (Brown, Askew, Baker, Denvir, & Millett, 1998, p. 378; Brown, Askew, Millett, & Rhodes, 2003, p. 670). Under these conditions, there was little alternative but to largely endorse a scaling up of the approach of the existing National Numeracy Project to provide the basis for the National Numeracy Strategy (Brown, Millett, Bibby, & Johnson, 2000, pp. 460–461). Hence, the Task Force Reports (DfEE, 1998a,b) were written in a style in which professional judgments from privileged sources were corroborated by general appeals to "the research literature", as illustrated by this opening paragraph from the brief section entitled "Whole class teaching":

> Inspection evidence and the experience of the National Numeracy Project point to an association between more successful teaching of numeracy and a higher proportion of whole class teaching. There is support for this in the research literature, which also identifies the quality of the teaching as the key factor. (DfEE, 1998a, p. 19)

The reports themselves contained few explicit references to research. Although a more substantial bibliography of what was represented as relevant research was eventually published (Reynolds & Muijs, 1999), no public account was offered of how consideration of this material had influenced the recommendations made by the Task Force. However, addressing the question, "Is the National Numeracy Strategy research-based?" Brown and colleagues were able to draw on direct experience of the deliberations of the Task Force in concluding that:

> [S]ometimes recommendations are quite strongly underpinned, sometimes the evidence is ambiguous, sometimes there is little relevant literature, and sometimes the research is at odds with the recommendations ... The research findings are sometimes equivocal and allow differences of interpretation. ... The complexity of the findings and of the possible interpretations suggests that ministerial desires for simply telling 'what works' are unrealistic. (Brown et al., 1998, p. 378)

Nevertheless, they (Brown et al., 2003, pp. 656–657) also note that once responsibility for implementation of the National Numeracy Strategy moved on from the Task Force, account was taken of ways in which insights from relevant research in mathematics education could guide the more detailed formulations required, particularly as regards the elaboration of learning objectives, curricular sequences and teaching approaches.

The key features of the classroom approach advocated in the Task Force Reports were an emphasis on calculation, especially mental calculation; adoption of a standard three-part template for daily mathematics lessons, incorporating direct interactive teaching of whole classes and groups; and meticulous planning of teaching based on a detailed framework of learning objectives linked to a weekly schedule. However, in keeping with the strong influence on the Task Force of ideas from the field of school effectiveness and improvement, other recommendations addressed those aspects of the reform with equal vigour. In particular, the Strategy showed no embarrassment about – in Bishop's terms – "perpetuat[ing a] centre-periphery model of educational change" (p. 195) and "assum[ing a] power structure which accords the researcher's agenda and actions greater authority than the practitioner's" (p. 192). However, the underlying model of change was more sophisticated in at least two important respects than the representation by Popkewitz as outlined by Bishop (p. 195). First, the various bodies of research and experience that the Strategy drew on included many directly concerned with practice. Second, the approach to leveraging change recognised the need to complement the "high pressure" of accountability already established within the system with correspondingly "high support" for schools and teachers in the form of much stronger incentives and mechanisms for building professional capacity for educational improvement (Earl et al., 2003: p. 130). Indeed, through this "high pressure and high support" approach the Strategy might be seen as taking account of Desforges and Cockburn's injunction – as quoted by Bishop – to accept responsibility for "provid[ing] the crucial psychological parameters of the teaching environment to which teachers and children alike must adapt" (p. 198).

While bent on changing teaching practice, however, the Task Force was alert to the need to take account of "teachers' behaviours in the context of ... teachers' reality" (p. 197). One consideration which weighed strongly with members of the Task Force in favour of adopting the National Numeracy Project as the basis for a national strategy was the positive evaluation it had received from teachers already involved in the project. These teachers identified certain features of the project's provision as particularly helpful: the detailed framework of learning objectives, the booklet of exemplar lessons, the training provided on mental calculation strategies, and in-school support from advisory teachers (Brown et al., 2000, p. 461). Contrary to the misgivings expressed by many academics, teachers across the country were similarly positive about the National Numeracy Strategy as its implementation developed (Brown et al., 2000; Earl et al., 2003). The external evaluation indicated that the Strategy's approach of seeking to provide high support to help teachers respond to the high pressure placed on them had resonated with practitioners. The evaluation reported that "schools were inclined to acquiesce to, and approve of" what had been termed the "informed prescription" of the Strategy, noting that "[h]eadteachers and teachers often expressed relief that they had been given the frameworks and curriculum materials to better cope with the pressure from national tests, Ofsted inspections, imposed targets and high workloads" (Earl et al., 2003, p. 130). For example, 74% of teachers agreed with statements to the effect that their teaching was more effective as a result of the Strategy, against 7% disagreeing; and 59% agreed that the Strategy had helped make their job more satisfying and engaging, against 12% disagreeing (Earl et al., 2003, pp. 85–86). Such approval was not confined to primary teachers; an external evaluation of the Strategy's extension to secondary schools reported similarly that mathematics teachers at that level generally found the framework of learning objectives very helpful, had positive reactions to the training provided, and valued the support from advisory teachers (Stoll et al., 2003, pp. 36–41).

The Redux

The external evaluation of the National Literacy and Numeracy Strategies reported that teachers were positive about the influence of the latter on aspects of pupil learning (Earl et al., 2003, p. 82). In the crucial terms of pupil performance on national tests at the end of primary school, the early years of the National Numeracy Strategy certainly saw a rise in the proportion of pupils achieving the benchmark level (Earl et al., 2003, p. 128). However, while gains were made until 2000, rates stalled in subsequent years, suggesting that a plateau had already been reached. Moreover, a study based on analysis of repeated administrations of an independent test-series to large national samples of pupils, conducted over the period from 1998 to 2002, showed a very modest rise in performance (Brown et al., 2003). Both major studies of the implementation of the Strategy suggested that early improvement of performance on the national tests resulted, to an important degree, from sharper focusing of classroom activity on their particular demands (Brown et al., 2003, p. 669; Earl et al., 2003, p. 137). Nevertheless, more specific ways were identified in which the treatment of particular aspects of mathematics had undergone widespread and beneficial change. The external evaluation reported that the majority of headteachers agreed that teachers in their school had significantly changed their teaching practices in mental mathematics as a result of the Strategy (Earl, Levin, Leithwood, Fullan, & Watson, 2001, p. 50). Correspondingly, the majority of teachers agreed that pupils were performing at a higher level in oral/mental mathematics as a result of the Strategy (Earl et al., 2003, p. 82). Moreover, in the other major study, fuller analysis at the item level of the evidence from independent testing indicated that "in general, those areas in which there is an improvement are those where it is clear that guidance given [by the Strategy] ha[s] updated the ways that topics have been taught in line with research findings, and increased the time allocated to them" (Brown et al., 2003, p. 667).

The key changes in classroom organisation and resources resulting from implementation of the Strategy are summarised in the external evaluation as follows:

> Up to the early or mid 1990s, schools were characterised by a predominance of individualised planning and teaching, with pace largely determined by pupil readiness as perceived by teachers. In mathematics, many teachers used commercial schemes of work, which children worked through at their own rate, often with little direct teacher intervention. The big shifts as a result of the Strateg[y] have been greater use of whole class teaching, greater attention to the pace of lessons, and planning based on objectives rather than activities. Most teachers are using the format and structure of ... the three-part daily mathematics lesson, although most have modified these as they gained confidence. (Earl et al., 2003, p. 127)

Overall, however, the external evaluation concluded that:

> Although the most obvious features of the reforms appear in virtually all classrooms, our data show considerable disparity across teachers and schools in understanding of the Strateg[y] and in subject and pedagogical knowledge and skill. In many cases the Strateg[y] ha[s] not yet produced the needed depth of change in teaching and learning. (Earl et al., 2003, p. 140)

The other major study focused more directly on change in teaching practice and beliefs and drew the similar conclusion that "teaching in the classroom seems to have changed mainly in superficial ways, e.g. organisation of lessons and resources used", whereas "[w]hen the beliefs of the teachers about how children should learn and be taught numeracy ..., and the way that teachers interact with children, are examined, it appears that in almost no cases have 'deep' changes taken place" (Brown et al., 2003, p. 668).

The reported prevalence of overly mechanical implementation of the Strategy suggests that it has been more successful in implanting a concrete apparatus of pedagogical practice in schools than in helping teachers to form this apparatus into a cogent system of tools which they are able to employ in flexible and discriminating ways. In secondary mathematics, for example, school inspections (OfStEd, 2004) indicate that while "the influence of the Strategy on curriculum planning has been beneficial where departments have revised their existing schemes of work to take account of the *Framework* and make best use of the available resources" (para. 79), this contrasts with the situation in around half the departments visited which have "adopted the sample medium-term plans issued by the Strategy too uncritically" (para. 80). Likewise, while "in the better practice, teachers use the recommended lesson structure in a flexible way, seeing it as a useful approach rather than a binding requirement" (para. 82), "in unsatisfactory lessons, teachers use a three-part lesson structure without thinking through the purpose of its various parts" (para. 84). Similarly, use by schools of internal assessment to focus on developing and securing pupil understanding of particular "key objectives" is contrasted favourably with use of short tests to award finely graded "attainment levels" (para. 88).

During the early years of the Strategy, schools and teachers were steered towards mechanical implementation through the strong direction from the centre and heavy pressure for compliance, which led to it being perceived as "a one-size-fits-all approach to teaching imposed on a widely diverse range of schools" (Earl et al., 2003. pp. 7, 135). While the small print of Strategy documents may have acknowledged possibilities of variation and adaptation, the bold titles of Strategy presentations conveyed a more prescriptive message, as did school inspections explicitly focused on recommended features from the Strategy (Stoll et al., 2003, pp. 31, 34). This inflexible approach represented an unpromising start for a policy that purported to build professional capacity for thoughtful adoption and localised adaptation; rather it was intended to meet the political imperative for rapid improvement in headline test performance. The external evaluation of the Strategy suggests that a shift towards a more devolved approach has become desirable as relatively straightforward initial gains have been exhausted. However, moving from a culture emphasising conformity to one encouraging local initiative through introducing greater flexibility over implementation creates a challenge of how "to push towards conditions where . . . schools and teachers have the capacity to adapt, solve problems and refine their practice, while remaining true to the principles underlying the Strategies" (Earl et al., 2003, p. 135). The reported differences in the degree to which schools and teachers have gone beyond superficial implementation of the Strategy to informed interpretation and adaptation appear to have been strongly influenced by existing variation in school and teacher capacity and confidence. The evaluation of the Strategy reports that many schools found it difficult to put such factors in place, let alone approach the ideal of the school as a learning organisation; it points to a considerable need for more sustained professional development aimed at promoting confident handling of subject matter and informed reflection on pedagogical issues. In effect, it constitutes recognition of the need for the new institutional structuring of educational improvement to be complemented by deeper processes of research-informed professional development of the type endorsed by Bishop.

Lessons to be Learned

This history of a quarter-century of systemic improvement effort in English mathematics education is, of course, something of a caricatural one. However, use of caricature as a didactic device aims to focus attention on key lessons.

The first of these key lessons was announced in my opening comments: that official agencies and policy-making bodies exert as important an influence on the researcher world as they do on the practitioner, shaping the kinds of educational research, developmental activity, and other forms of researcher engagement with practice which can be undertaken. This has been an explicit thread of the argument, highlighted by the way in which national policy shifts during the late 1980s radically changed the terms on which university-based mathematics educators participated in initial teacher education and continuing professional development, marginalising the reflective dimension of these programmes, and limiting scope for the kinds of developmental research in which many university mathematics educators had been involved during the post-Cockcroft era. However, the new emphasis on educational evaluation and systemic change created opportunities for other fields of educational research, notably that of school effectiveness and improvement, which was better able to engage with the new political agenda on its own terms. We can see this shift clearly if we compare the briefs of the Cockcroft Committee and the Numeracy Task Force. Whereas the Cockcroft Committee was asked "to consider the teaching of mathematics ... with particular regard to its effectiveness and intelligibility and to the match between the mathematical curriculum and the skills required in further education, employment and adult life generally" (Cockcroft, 1982, p. ix), the remit of the Numeracy Task Force "to develop a national strategy to raise standards of numeracy in order to reach the national numeracy target by 2002" (DfEE, 1988a, p. 4) was much more technocratically framed and narrowly focused. Equally, whereas the Cockcroft Committee adopted a problematic largely drawn from the field of mathematics education, the Numeracy Task Force was more strongly influenced by conceptualisations developed within the fields of teacher effectiveness and school improvement. Finally, in contrast to the lengthy period granted the Cockcroft Committee for its deliberations (stretching from 1978 to 1982), which enabled it to carry out extensive groundwork and build professional and political consensus, the Numeracy Task Force was pressed to produce recommendations in line with the new government's policy aspirations within six months, and – following an interval for consultation – an implementation strategy within an even shorter period.

To the degree that, as Bishop suggests, "[m]ost [researchers] would probably still yearn longingly for the academic ideal of the disinterested researcher" (p. 198), it would be easy to denigrate the shift in political culture between the Cockcroft Report and the Numeracy Task Force. However, fuller reflection shows that this shift raises some uncomfortable questions which need to be considered by the mathematics education community as a whole (not just by the researchers within it). First, mathematics education is by definition selective in its attention. It develops powerful understandings of the specifically mathematical aspects of educational practice, but it either neglects many other aspects of this practice (such as many on which the school effectiveness and improvement field focuses), or conceptualises them in ways which prioritise the mathematically significant dimensions over others. The result is that mathematics education alone is often unable to adequately address holistic problems of policy and practice of the types which the Cockcroft Committee and the Numeracy Task Force were established to tackle. Equally, the understandable valorisation of mathematical aspects over others has had the unfortunate corollary of encouraging an insularity which poorly equips the field to enter the kinds of collaboration necessary to address many issues of educational policy and practice through bringing the specialised perspectives of several fields to bear on them. Successful collaboration of this type has to be more than simply an *ad hoc* matter. While lengthy deliberations and full consultations with different constituencies enabled the Cockcroft Committee to establish wide professional and political support, the intellectual basis on which that support was gained was not consolidated sufficiently and developed adequately to meet the challenges that emerged over subsequent years. For example, wider support for a conception of mathematical capability which extended beyond "basic skills" and "simple tests" to encompass relational understanding, realistic application, and technology integration proved relatively short-lived. In particular, under the pressured conditions in which the Numeracy Task Force operated, the absence of established common ground between the school effectiveness and mathematics education specialists had far reaching consequences. For example, not only did it act as a significant impediment to securing support for that broader conception of mathematical capability, it led to much greater weight being attached to the pedagogical models endorsed by earlier teacher effectiveness research, which took narrower definitions of mathematical capability for granted. This reveals a second key lesson: that mathematics education alone often provides insufficient resources to formulate adequate responses to issues of policy and practice; and thus that mathematics education needs to develop stronger interactions with complementary professional fields. Perhaps this is a development for which a future ICMI conference on Cooperation between Research Fields in the Systemic Improvement of Mathematics Education could provide international leadership?

These key lessons accord, then, with the spirit of Bishop's closing recommendations that "[i]nstitutional context and constraints should be given greater prominence in research" and that "[t]he process of change needs to be a greater focus in research on mathematics teaching" (p. 199), but to suggest that the most effective strategy might be for mathematics education to build stronger collaboration and dialogue with those fields of educational enquiry and practice which focus specifically on such issues. This leads to a third key lesson: that it is important that mathematics education retain its clear focus but develop its capacity for synergy with wider perspectives. This includes a broadening of Bishop's recommendation that "[t]eam research by researchers/practitioners should be emphasised" (p. 199) to encompass greater multidisciplinarity of researchers within such teams. Huberman has pointed to some of the benefits of "sustained interaction" between researchers and teachers, "in which researchers defend their findings and some practitioners dismiss them, transform them, or use them selectively and strategically in their own settings" (Huberman, 1993, p. 34). Reframing ideas in order to address the qualifications and challenges identified through collaboration with teachers appears to trigger a decentring process amongst researchers. In particular, it creates a need to marshal a broader range of scholarly thinking and research experience; and teams containing mathematics educators alone are often ill equipped to do this. But multidisciplinary teams do not simply enrich the range of ideas and techniques available; by introducing further perspectives and fresh challenges, they help mathematics education researchers to refine their tools and strengthen their arguments. Strong collaboration with teachers is also important in research aiming to develop professional practice because of the person-embodied, tool-mediated and setting-embedded practitioner craft that is key to making such practice realisable. In this respect I would strengthen Bishop's recommendation that "Researchers need to focus more attention on practitioners' everyday situations and perspectives" (p. 199), to argue that developmental research needs to incorporate a dialogic cycle of knowledge-creation through which, on the one hand, theorised scholarly knowledge is contextualised and activated within teaching, stimulating construction of relevant practitioner craft knowledge; while, on the other hand, some of that teacher craft knowledge is elicited and codified in a form which can help improve the effectiveness with which the training and coaching of other teachers can be undertaken, by providing more explicit frameworks for analysing teaching processes, for articulating mechanisms and functions, and for understanding adaptation to different conditions (Ruthven, 2002).

The last of Bishop's recommendations deals with the dissemination of research. In comparison with the efforts expended on this over both the post-Cockcroft and post-ERA periods, and the highly active approaches adopted, the recommendation that "[c]onclusions and outcomes should be published in forms which are accessible to the maximum number of practitioners" (p. 199) seems an unambitious recommendation from a paper seeking to encourage researchers to enter and engage with the practitioners' world. However, it could be seen as an appropriately modest response to the thread of disappointment with the effectiveness of dissemination which runs through this history. Indeed, this disappointment might seem to oblige us to accept Desforges and Cockburn's pessimistic conclusion – as quoted and endorsed by Bishop – that "classrooms as presently conceived and resourced are simply not good places in which to expect the development of the sorts of higher-order skills currently desired from a mathematics curriculum" (p. 199). But the proponents of "reflective" and "effective" practice would both counter with powerful examples of schools and teachers where the approach in question has supported improvements in practice. Rather, a more plausible reading of this history is that each approach has been more successful where important elements of the other were also present: practitioner-centred reflective practice disciplined by careful attention to clear objectives and hard evidence; research-informed effective practice mediated by educated use and thoughtful adaptation of the techniques and tools provided. Indeed, this was the opinion of two of the doyens of the effective teaching field from whose work – ironically – much of the National Numeracy Strategy's prescription for effective practice was derived:

> It should also be stressed that there is no single system for presenting mathematics concepts effectively. For example, some of the control teachers in our studies have obtained high levels of student achievement using instructional systems that differed from those in the program we developed ... Research ... yields ... concepts and criteria that can be applied in order to examine classroom instruction. Hence, in current work [we] are not advocating

> that teachers ... implement behaviour in a mechanical fashion. Rather, [our] efforts are to use the findings to stimulate teachers to discuss ... the various ways this model can be implemented in classrooms. (Good & Biddle, 1988, p. 131).

The last lesson, then, is that success in developing research-informed practice is likely to depend both on dialogic development of scholarly and craft knowledge at the stage of research, and on productive interplay between "reflective" and "effective" approaches at the stage of wider dissemination. Happily, this is a lesson that the English government appears now to be starting to learn. It has recently supported the establishment of a National Centre for Excellence in Teaching Mathematics on a model which recognises that building professional capacity in the field depends on processes of both these types.

References

Askew, M., & Wiliam, D. (1995). *Recent Research in Mathematics Education 5–16*. London: HMSO.

Bishop, A. (1991). Teaching mathematics to ethnic minority pupils in secondary schools. In D. Pimm & E. Love (Eds.), *Teaching and learning school mathematics* (pp. 26–43). London: Hodder & Stoughton.

Bishop, A. (1998). Research, effectiveness, and the practitioners' world. In A. Sierpinska & J. Kilpatrick (Eds.), *Mathematics education as a research domain: A search for identity* (pp. 33–45). Kluwer: Dordrecht.

Bishop, A., & Nickson, M. (1983). *A Review of research in mathematical education, Part B, research on the social context of mathematics education*. Windsor: NFER-Nelson.

Brown, M. (1993). Clashing epistemologies: The battle for control of the national curriculum and its assessment. *Teaching Mathematics and its Applications, 12*(3), 97–112.

Brown, M., Askew, M., Baker, D., Denvir, H., & Millett, A. (1998). Is the national numeracy strategy research-based? *British Journal of Educational Studies, 46*(4), 362–385.

Brown, M., Millett, A., Bibby, T., & Johnson, D.C. (2000). Turning our attention from the what to the how: The national numeracy strategy. *British Educational Research Journal, 26*(4), 457–471.

Brown, M., Askew, M., Millett, A., & Rhodes, V. (2003). The key role of educational research in the development and evaluation of the national numeracy strategy. *British Educational Research Journal, 29*(5), 655–672.

Cockcroft, W. H. (Ch.) (1982). *Mathematics counts*. Report of the Committee of Inquiry into the Teaching of Mathematics in Schools. London: HMSO.

Department for Education and Employment [DfEE] (1998a). *Numeracy matters: The preliminary report of the numeracy task force*. London: DfEE.

Department for Education and Employment [DfEE] (1998b). *The implementation of the national numeracy strategy: The final report of the numeracy task force*. London: DfEE.

Dowling, P., & Noss, R. (Eds.) (1990). *Mathematics versus the national curriculum*. London: Falmer.

Earl, L., Levin, B., Leithwood, K., Fullan, M., & Watson, N. (2001). *Watching and learning 2: OISE/UT evaluation of the implementation of the National Literacy and Numeracy Strategies*. Toronto: OISE/UT.

Earl, L., Watson, N., Levin, B., Leithwood, K., Fullan, M., & Torrance, N. (2003). *Watching and learning 3: Final report of the external evaluation of England's National Literacy and National Numeracy Strategies*. Toronto: OISE/UT.

Elliott, J., & Sarland, C. (1995). A study of 'teachers as researchers' in the context of award-bearing courses and research degrees. *British Educational Research Journal, 21*(3), 371–386.

Elliott, J., Maclure, M., & Sarland, C. (1997). Teachers as researchers in the context of award bearing courses and research degrees: Summary of research results. Centre for Applied Research in Education, University of East Anglia. Retrieved at http://www.uea.ac.uk/care/research/tar.html on 31/05/05.

Good, T. L., & Biddle, B. J. (1988). Research and the improvement of mathematics instruction: The need for observational resources. In D. Grouws, T. J. Cooney, & D. Jones (Eds.), *Effective mathematics teaching* (pp. 114–142). Reston VA: Lawrence Erlbaum Associates/National Council of Teachers of Mathematics.

Huberman, M. (1993). Changing minds: The dissemination of research and its effects on practice and theory. In C. Day, J. Calderhead & P. Denicolo (Eds.), *Research on teacher thinking: Understanding professional development* (pp. 34–52). London: Falmer.

Low Attainers in Mathematics Project [LAMP] (1987). *Better mathematics: A curriculum development study*. London: HMSO.

McIntyre, D. (1995). Initial teacher education as practical theorising: a response to Paul Hirst. *British Journal of Educational Studies, 43*(4), 365–383.

Office for Standards in Education [OfStEd] (2004). *The key stage 3 strategy: Evaluation of the third year*. London: OfStEd.

Raising Achievement in Mathematics Project [RAMP] (1991). *Raising achievement in mathematics project 1986–89: A curriculum development study*. Bognor: The Mathematics Centre, West Sussex Institute of Higher Education.

Reynolds, D., & Muijs, D. (1999). *National Numeracy Strategy: an annotated bibliography for teachers and schools*. London: DfEE.

Ruthven, K. (1989). An exploratory approach to advanced mathematics. *Educational Studies in Mathematics, 20*(4), 449–467.

Ruthven, K. (1999). Reconstructing professional judgement in mathematics education: From good practice to warranted practice. In C. Hoyles, C. Morgan & G. Woodhouse (Eds.) *Rethinking the mathematics curriculum* (pp. 203–216). London: Falmer.

Ruthven, K. (2001). Mathematics teaching, teacher education and educational research: developing 'practical theorising' in initial teacher education. In F.-L. Lin & T. Cooney (Eds.) *Making sense of mathematics teacher education* (pp. 165–183). Dordrecht: Kluwer.

Ruthven, K. (2002). Linking researching with teaching: towards synergy of scholarly and craft knowledge. In L. English (Ed.) *Handbook of international research in mathematics education* (pp. 581–598). Mahwah NJ: Lawrence Erlbaum Associates.

Ruthven, K. (2005). Improving the development and warranting of good practice in teaching. *Cambridge Journal of Education*, 35(3), 407–426.

Sierpinska, A., & Kilpatrick, J. (Eds.) (1998). *Mathematics education as a research domain: A search for identity*. Dordrecht: Kluwer.

Stoll, L., Stobart, G., Martin, S., Freeman, S., Freedman, E., Sammons, P., & Smees, R. (2003). *Preparing for change: Evaluation of the implementation of the Key Stage 3 Strategy pilot*. London: DfES.

Section VII
Values

Intuitively many researchers have believed that affect impacts on learning. Bishop was no different. Although others had concentrated on student affect, Bishop in his thinking about how teachers should teach, through the 1970s recognised that teachers' attitudes, beliefs and values had some role to play. He included these in the diagram used for his teacher decision-making research (see Fig. 1, introduction to Section 2). By the late 1980s he began to focus on just what these notions were, of beliefs and values as they pertained to mathematics, and in mathematics teaching, and indeed in mathematics learning. By the mid 1990s his attention turned quite specifically to these ideas, and in particular how teachers taught various values in mathematics sessions.

Bishop by this time was at Monash University, Australia. The discussion within an interested group of researchers focused on how classrooms could be probed to gain some understanding of whether teachers actually recognised that they taught values, what values they did teach, and whether teachers could change the way they taught to emphasize in their teaching identified values they may not have been teaching previously. And so the *Values and Mathematics Project* (VAMP) began. Needless to say many teachers then, and still today, respond when initially asked as to what values they teach in mathematics sessions with "None. I'm teaching mathematics." At the heart of the observational studies that VAMP undertook was trying to understand what values were playing a role when teachers were making critical decisions in the flow of activity of their mathematics teaching: thus there was a return to Bishop's 1970's notion.

As Clarkson (2007) has commented elsewhere, in reference to this section's key article:

> In this article Bishop goes back to the literature to give meaning to the term values. In later work the operational meaning of the term grows from the meanings gained from observational studies. The key notion that emerges is that students do carefully observe their teacher, and they deduce from the teacher's behaviours what the teacher values. These are what students react to. Hence although the teacher may believe whatever, it is their behaviours in the classroom that define what they value in that context.
>
> This article was written when there was renewed interest in society in values. As the economic rationalists secured their dominant position in politics, interestingly there was a divergent discussion in society that gathered momentum. Was everything to be judged in economic terms? Was there no room for other than judgements made against measurable,

> 'objective' criteria? The divergent discussion suggested yes, there were other values that are inherent in the cultural artefacts that we have created. As with other artefacts, this article began a debate on what were the implied values inherent in mathematics and the pedagogy of mathematics that we need to know about if we are to teach our students well. (p. 223)

Seah, who was one of the members of the VAMP project team, had been undertaking parallel research projects for initially his master's thesis and then his doctoral thesis under the supervision of Bishop. In his contribution to this section he explores where the research that started with Bishop has led him in his own studies, and in so doing comments on the deeper understandings of the terms we now use, compared to what in retrospect appears to be quite limited understanding of these just ten years ago.

During the life of the VAMP project, contact was made and retained with a parallel group working in Taiwan led by Fou-Lai Lin and Chien Chin, both of whom had undertaken research studies at Cambridge University under the supervision of Bishop. Members of the VAMP team individually (Bishop once and Clarkson twice), and then as a group just before the History and Pedagogy of Mathematics 2000 Conference (see Horng & Lin, 2000), visited Taiwan and developed an important working relationship with that group. It was challenging to the Australian group, including Bishop, to rethink ideas and try and take into account stories told of classroom interactions of Taiwanese mathematics teachers, teaching obviously within an ethnic Chinese culture, and some of whom had strong Confucian or Buddhist beliefs. Seah references some of the Taiwanese group's publications that were important to Bishop's continuing thinking in this area.

An Additional Bishop Reference Pertinent to this Issue

Bishop, A. J. (1991). Mathematical values in the teaching process. In A. J. Bishop, J. van Dormolen, & S. Mellin-Olsen (Eds.), *Mathematical knowledge: Its growth through teaching*. Dordrecht: Kluwer Academic Publishers.

References

Clarkson, P. C. (2007). The development of research into values in mathematics education. In G. Leder & H. Forgasz (Eds.), *Stepping stones for the 21st century* (p. 223). Rotterdam: Sense Publishers.

Horng, W. S., & Lin, F. -L. (Eds.) (2000). *Proceedings of the History and Philosophy of Mathematics Conference: History in mathematics education – Challenges for a new millennium*. Taipei: National Taiwan Normal University.

Chapter 16
Mathematics Teaching and Values Education – An Intersection in Need of Research

Alan J. Bishop, Melbourne (Australia)

Kurzreferat: *Mathematikunterricht und Werteerziehung – eine Schnittstelle, die der Forschung bedarf.* In diesem Beitrag werden Fragen zur Werteerziehung im Rahmen des Mathematikunterrichts aufgeworfen. Es wird argumentiert, daß Erziehung zur Demokratie immer auch Werteerziehung miteinschließt. Es wird ein überblick über relevante Forschungsarbeiten und theoretische Perspektiven gegeben und für verstärkte Forschungsaktivitäten in diesem Bereich plädiert. Obwohl es relevante Forschung im affektiven Bereich gibt, sowohl allgemein wie auch auf Mathematik bezogen, im sozialen und im kulturellen Kontext, so gibt es wenig direkte Forschung, die sich auf Werte oder Wertsetzung bezieht. Lehrer sind sich selten darüber bewußt, daß sie implizit oder explizit Werte vermitteln, aber Wertevermittlung findet statt, und zwar meistens implizit. Wenn der Mathematikunterricht dahingehend verändert und verbessert werden soll, daß er dem Leben in modernen demokratischen Gesellschaften gerechter wird, dann muß zunächst dieser Aspekt der mathematischen Erziehung besser verstanden werden.

Introduction and Context

Modern society is demanding much greater mathematical knowledge of its citizens than ever before and the essential challenge for mathematics educators concerned with issues of democracy is how to provide an adequate mathematics education for the greatest number of citizens. Computer developments are simultaneously facing us with some of our greatest dilemmas, and offering us some of the most exciting educational possibilities. They are not only changing the way we think about mathematics teaching, they are also changing the nature of mathematical activity itself. Societies are also becoming more multi-cultural and because the nature of mathematics is being re-examined, mathematics educators are becoming increasingly concerned about the goals which should be formulated for mathematics education. There are important developments in the last few years which could have widespread benefits for mathematics learners around the world, for example in the areas of ethnomathematics (see Gerdes, 1995) and critical mathematics education (see Skovsmose, 1994, 1996) where both the nature and the role of mathematics in education is being looked at anew.

Mathematics Teaching and Democratic Education. Part 2 ZDM 31(1999) 1–4 1863-9690.

What is particular provocative about this general situation and these examples is that there is a strong interest in examining, and in changing, the values being taught through mathematics education. But only rarely does one find explicit values teaching going on in mathematics classrooms. Why would this be? The reason lies in the widespread belief that mathematics is a value-free subject, a myth which has been exploded in the last two decades.

Dealing with issues of democracy in mathematics education clearly requires engaging with values, and this is problematic at the present time because not only do we not know what currently happens with values teaching in mathematics classrooms, or why, but we have even less idea of how potentially controllable such values teaching is by teachers. In addition, whereas it appears to be relatively easy in the teaching of humanities and arts subjects to record, discuss and develop values teaching and learning, this is not the case at present in mathematics teaching. Most mathematics teachers would not even consider that they are teaching any values when they teach mathematics. Changing that perception may prove to be one of the biggest hurdles to be overcome.

There are therefore several important questions which are worth considering here: What is the current situation regarding values teaching in mathematics classrooms? What values do mathematics teachers think they are teaching? What values are being learnt by students? Can teachers gain sufficient control over their values teaching to teach other values besides those which they currently teach? Sadly there is very little research into any of these questions, which is creating a huge lacuna in our understanding of how to affect the current situation. Before discussing those questions in more detail, however, it is necessary to clarify what we mean when we talk about values in mathematics education.

Values and Mathematics Education: Three Sources of Conceptualisations

Values in mathematics education are the deep affective qualities which education fosters through the school subject of mathematics. They appear to survive longer in peoples' memories than does conceptual and procedural knowledge, which unless it is regularly used tends to fade. Research indicates that the negative features of these values lead subsequently to a dislike of mathematics as adults and hence to negative parental influence (Cockcroft, 1982).

If we consider the relevant research fields, we can find three principal sources for theoretical ideas which can be used to think about developing values teaching in mathematics. These are the literatures on the affective domain and values education generally, on affective aspects of mathematics education, and on social and cultural aspects of mathematics education.

Affective Domain and Values Education

The first framework offered to address these issues was Krathwohl, Bloom and Masia's (1964) analysis of the affective domain of the taxonomy of educational objectives, which introduced the difference between "values" and "valuing". Their analysis suggested five levels of response to a phenomenon in increasing degrees of commitment. Of particular interest here are levels 3 and 4 which are summarised as follows:

3. Valuing
3.1 acceptance of value
3.2 preference for a value
3.3 commitment
4. Organization
4.1 conceptualization of a value
4.2 organization of a value system.

The following quotation is important for clarifying the focus of our concerns:

> Behavior categorized at this level is sufficiently consistent and stable to have taken on the characteristics of a belief or an attitude. The learner displays this behavior with sufficient consistency in appropriate situations that he comes to be perceived as holding a value. (p. 180)

Raths, Harmin and Simon (1987), summarising their often-quoted book, approach the problem in another way, and offer seven criteria for calling something a value. They say (p. 199): "Unless something satisfies *all* seven of the criteria noted below, we do not call it a value, but rather a 'belief' or 'attitude' or something other than a value." They summarise their criteria in the following terms:

1. Choosing freely
2. Choosing from alternatives
3. Choosing after thoughtful consideration of the consequences of each alternative
4. Prizing and cherishing
5. Affirming
6. Acting upon choices
7. Repeating.

They add "Those processes collectively define valuing. Results of this valuing process are called values" (p. 201).

Both the taxonomy and the criteria from Raths et al. emphasise the following aspects of valuing which are important for our consideration:

(i) the existence of alternatives
(ii) choices and choosing
(iii) preferences
(iv) consistency.

In relation to values education, the work of Tomlinson and Quinton (1986) is particularly important since it moves the discussion from earlier reliance on the work of Kohlberg (1984) and his followers into the mainstream subject curriculum. They argue strongly that when considering values, due attention should be paid to three elements (p. 3): aims or intended outcomes; means or teaching/learning processes; and effects or actual outcomes. This same triad shaped the work of the IEA comparative research on mathematics teaching (see Garden, 1987) which focused attention on three levels of the curriculum: the intended level, the implemented level, and the attained level. These are clearly important ideas for us to consider here.

Affective Aspects of Mathematics Education

Regarding the second literature source, McLeod (1992), in one of the most up-to-date and comprehensive summaries of research into affective aspects of mathematics education, separates the field into studies of beliefs, attitudes, and emotions. He, like others who have surveyed this field, cites no research on values, although the tone of his discussion makes it clear that, rather like Krathwowhl et al. and Raths et al. above, ideas about both beliefs and attitudes towards mathematics do relate to values held by both teachers and students.

In another chapter in the same book Thompson (1992) also discusses the research on teacher beliefs, particularly in relation to teachers' actions in the classroom. She points to a repeated finding that teachers' actions frequently bore no relation to their professed beliefs about mathematics and mathematics teaching. The research by Sosniak, Ethington and Varelas (1991) also found striking inconsistencies between different belief statements given by the same teachers. We would contend that this discrepancy is precisely why it is necessary to focus on values rather than beliefs, in order to determine the deeper affective qualities that are likely to underpin teachers' preferred decisions and actions.

Taking into consideration the findings from other research on beliefs and attitudes in mathematics education (e.g. Buxton, 1981, Fasheh, 1982), of more concern is the fact that there appear to have been few studies of the behavioural aspects of affect, such as those behaviours described earlier related to valuing – namely, the choosing, the preferring, the consistency of behaviour, etc. The behavioural component certainly appears to be one significant focus for the development of attitudes and beliefs on the one hand, and values on the other.

Social and Cultural Aspects of Mathematics Education

This third literature source has been helpful in clarifying what is the *content* or the *focus* of the values which should be addressed. As was stated in the opening paragraph of this paper, there are three principal sources of values in the mathematics classroom; society, mathematics, and mathematics education.

Wilson's (1986) review, whilst pointing out the paucity of writing and research on values in mathematics teaching did discuss two values, *a respect for truth*, and *the authority of mathematics*. Later analyses by Bishop (1988 and 1991) sought to build more broadly on the wide literature on mathematical history and culture. Using White's (1959) three component analysis and terminology, he proposed that, in "Western" mathematics development, the predominant ideological values concern the ideas of "rationalism" and "objectism", the sentimental values (which is White's term for individuals' feelings about their relationship to knowledge) are those of "control" and "progress", while the sociological values refer to societal relationships regarding mathematical knowledge, such as "openness" and "mystery". Wilson's (1986) first value is an ideological one, while the second fits comfortably within White's "sentimental" component.

It seems therefore that the three conceptualisations which will be important for values research to consider in the future are the following:

- Ideological: referring to the values of individuals towards the mathematics they are either teaching or learning
- Individual: referring to the values of individuals towards themselves, their self-respect etc. in the context of learning or teaching mathematics
- Social: referring to the values of individuals towards society, in relation to mathematics education.

The Teacher and Values Education in Mathematics

As was said above, only rarely does one find explicit values teaching going on in mathematics classrooms, the reason being the widespread belief that mathematics is a value-free subject. Indeed, many parents and politicians might initially be concerned about explicit values teaching in mathematics. What parents and others should be concerned about is that values teaching and learning do go on in mathematics classrooms, and because most of it appears to be done implicitly, there is only a limited understanding at present of what values are being transmitted, and of how effectively they are being transmitted. Given the often-quoted negative views expressed by adults about their bad mathematics learning experiences, one could speculate that the values transmitted to them were not necessarily the most desirable, but that they were transmitted rather effectively!

At present we have no research which is documenting the extent of values teaching. We have no idea what either the explicit or the implicit forms of values teaching are. Several questions come to mind here: Are values *explicitly* expounded, discussed or raised as teaching "content"? As they do not appear in detailed syllabus descriptions (Howson, 1991) but only, if at all, in the aims statements of curriculum documents, it is unlikely that they will be considered as content to be taught. The assumption will therefore be that they will be addressed (if at all) across and through the mathematical content or process topics.

Do textbooks have explicit values-focused exercises or activities? A look at several textbooks fails to reveal any activities of this nature, and again one would suspect that as values do not appear as content they would not be addressed by classroom texts. Do teachers use values clarification exercises, etc.?

From a research perspective the International Handbook on Mathematics Education (Bishop et al., 1996) is revealing. It has no specific chapter on values, although several of the chapters clearly refer to value aspects of mathematics education, and stress their importance. For example, Brown (1996) discusses the work of the Humanistic Mathematics Network and quotes one of its aims which states: "An understanding of the value judgements implied in the growth of any discipline. Logic alone never completely accounts for *what* is investigated, *how* it is investigated and *why* it is investigated" (p. 1302). Ernest (1996) also implicity discusses values in his chapter on "Popularization: myths, massmedia amd modernism" as do Leder et al. (1996) in the chapter on gender issues.

Skovsmose's (1996) chapter is perhaps the one which most nearly addresses values and valuing explicitly, when he argues that

> Critical mathematics education is concerned with the development of citizens who are able to take part in discussions and are able to make their own decisions. We therefore have to take into consideration the fact that students will also want, and should be given the opportunity, to 'evaluate' what happens in the classroom. This turns the focus on students' interest. (p. 1267)

This comment echoes the idea above, that for values education to develop there is a necessity to ensure that the mathematics classroom is a place of choices, and of choosing, for the students. Teachers could, and in my view should, be presenting students with activities which encourage them to make choices; for example, about the selection of problems to be solved; about the solution approaches to be taken; about the criteria for judging the worth of solutions; and about the wider appropriateness of the mathematical models being taught. It should be a natural part of the teacher's repertoire, to present activities which require choices to be made: for example, a task such as "Describe and compare three different proofs of the Pythagorean theorem" would inevitably engage students in discussing the values associated with proving. Even the simple act of presenting different problem-solving solutions to be compared and contrasted by the students stimulates the ideas of choice, criteria, and values. What Skovsmose's focus on students' interests does is to remind us that rather than thinking of mathematics teaching as just teaching mathematics to students, we are also teaching students through mathematics. They *are* learning values through how they are being taught.

This is also why more attitude-focused research needs to be refocused onto values, and choices. We need studies which not only investigate what students *say* about their attitudes to different aspects of mathematics, but also which look at the choices students make in different situations, which will indicate the influence of certain values.

The acceptability of these ideas will of course depend ultimately on the capacity of teachers to engage with this issue. For example, when choices are offered to the students and made by them, how do teachers respond? Do in fact teachers know what values they are currently implicitly teaching in the ways they respond to students? Are they in that sense in control of their own values teaching? These are of course crucial questions. Many development projects are predicated on the assumption that teachers *are* in control of their values teaching and that they *will* be able to change the values to which they teach. However it is largely an unexplored area. Perhaps only when teachers give students more choices will they themselves be faced with responses which are new to them, and which will therefore require them to become more aware of their own values. Perhaps indeed this is another inhibiting factor in the process: perhaps one reason mathematics teachers do not give their students more opportunities for choice is precisely because it will require them to examine and reveal the values about which they themselves are unsure.

This area is one which is fundamental not just to research, but also to much teacher training and in-service education, and it needs to be thoroughly investigated by both teachers and researchers. The results of any such in vestigations would do much to enlarge our understandings of why mathematics teachers teach in the ways they do, of how to educate mathematically our future citizens, and of what are desirable, and feasible, goals for mathematics education in democratic societies as we move towards the next century.

An Additional Bishop Reference Pertinent to this Issue

Bishop, A. J. (1999). Mathematics teaching and values education: An intersection in need of research. *Zentralblatt fuer Didaktik der Mathematik, 31*(1), 1–4.

References

Bishop, A. J. (1988): Mathematical enculturation: a cultural perspective on mathematics education. Dordrecht: Kluwer

Bishop, A. J. (1991): Mathematical values in the teaching process. In: A. J. Bishop; S. Mellin-Olsen; J. van Dormolen (Eds.), Mathematical knowledge: its growth through teaching. Dordrecht: Kluwer. p. 195–214

Bishop, A. J.; Clements, K.; Keitel, C.; Kilpatrick, J.; Laborde, C. (1996): International handbook of mathematics education. Dordrecht: Kluwer

Brown, S. I. (1996): Towards humanistic mathematics education. In: A. J. Bishop; K. Clements; C. Keitel; J. Kilpatrick; C. Laborde (1996), International handbook of mathematics education. Dordrecht: Kluwer, p. 1289–1321

Buxton, L. (1981): Do you panic about Maths? London: Heinemann

Cockcroft, W.H. (1982): Report of the Committee of Inquiry into the teaching of mathematics in schools. London: HMSO

Ernest, P. (1996): Popularization: myths, massmedia and modernism. In: A. J. Bishop; K. Clements; C. Keitel; J. Kilpatrick; C. Laborde (1996), International handbook of mathematics education. Dordrecht: Kluwer. p. 785–817

Fasheh, M. (1982): Mathematics, culture, and authority. In: For the Learning of Mathematics 3 (No. 2), p. 2–8

Garden, R. A. (1987): The second IEA mathematics study. In: Comparative Education Review 31 (No. 1), p. 47–68

Gerdes, P. (1995): Ethnomathematics and education in Africa. Sweden: Stockholm University

Howson, A. G. (1991): National curicula in mathematics. Leicester, UK.: The Mathematical Association

Kohlberg, L. (1984): Essays on moral development: the psychology of moral development (Vol. 2). New York: Harper and Row

Krathwohl, D. R.; Bloom, B.S.; Masia, B. B. (1964): Taxonomy of educational objectives, the classification of educational goals: Handbook 2: Affective Domain. New York: Longman

Leder, G. C.; Forgasz, H. J.; Solar, C. (1996): Research and intervention programs in mathematics education: a gendered issue. In: A. J. Bishop; K. Clements; C. Keitel; J. Kilpatrick; C. Laborde (1996), International handbook of mathematics education. Dordrecht: Kluwer, p. 198–214

McLeod, D. B. (1992): Research on affect in mathematics education: a reconceptualization. In: D. A. Grouws (Ed.), Handbook of research on mathematics teaching and learning. New York: Macmillan, p. 575–596

Raths, L. E.; Harmin, M.; Simon, S. B. (1987): Selections from values and teaching. In: P. F. Carbone (Ed.), Value theory and education. Malabar, USA: Krieger, p. 198–214

Skovsmose, O. (1994): Towards a philosophy of critical mathematical education. Dordrecht: Kluwer

Skovsmose, O. (1996): Critical mathematics education. In: A. J. Bishop; K. Clements; C. Keitel; J. Kilpatrick; C. Laborde (1996), International handbook of mathematics education. Dordrecht: Kluwer. p. 1257–1288

Sosniak, L.A.; Ethington, C. A.; Varelas, M. (1991): Teaching mathematics without a coherent point of view: findings from the IEA Second International Mathematics Study. In: Journal of Curriculum Studies 23 (No. 2), p. 119–131

Thompson, A. (1992): Teachers' beliefs and conceptions: a synthesis of the research. In: D. A. Grouws (Ed.), Handbook of research on mathematics teaching and learning. New York: Macmillan, p. 127–146

Tomlinson, P.; Quinton, M. (Eds.) (1986): Values across the curriculum. Lewes: The Falmer Press

White, L. A. (1959): The evolution of culture. New York: McGraw-Hill

Wilson, B. J. (1986): Values in mathematic education. In: P. Tomlinson; & M. Quinton (Eds.), Values across the curriculum. Lewes: The Falmer Press, p. 94–108

Chapter 17
Valuing Values in Mathematics Education

Wee Tiong Seah

Introduction

The research field of values in school mathematics teaching and learning has been conceptualised in explicit ways and developed since the late 1980s by Alan Bishop. The article that stimulated the composition of this chapter (see Bishop, 1999) is a demonstration of Bishop's identification of the need for the academic community to bring together considerations of mathematics teaching and values education if we are to teach mathematics successfully for democracy. Indeed, in that article, Bishop argued that both considerations of education for democracy and of making school mathematics more relevant to the demands of everyday living involve the teaching and inculcation of values to students. In Bishop's opinion,

> values in mathematics education are the deep affective qualities which education fosters through the school subject of mathematics. They appear to survive longer in people's memories than does conceptual and procedural knowledge, which unless it is regularly used tends to fade. (Bishop, 1999, p. 2)

As alluded to above, Bishop's interest in the values dimension of mathematics education can be traced back another decade or so, when he proposed a set of six mathematical values in his book *Mathematical enculturation: A cultural perspective on mathematics education* (Bishop, 1988). These six values were categorised into three complementary pairs, reflecting White's (1959) influence on his conception. White had earlier identified cultural growth as being driven by technology through the latter's stimulation along ideological, sentimental and sociological dimensions. In Bishop's conception, mathematics is one of these technologies, and cultural growth is facilitated through mathematics and other technologies, in turn through the three complementary pairs of values, each representing one of White's three dimensions listed above.

In this context, this chapter aims to take stock of some of the developments (and barriers to further development) of this aspect of mathematics education research

W.T. Seah
Faculty of Education, Monash University (Peninsula Campus), PO Box 527, Frankston, Vic. 3199, Australia
e-mail: WeeTiong.Seah@Education.monash.edu.au

P. Clarkson, N. Presmeg (eds.), *Critical Issues in Mathematics Education*,

and practice over the last ten years since the publication of Bishop's (1999) article, an exercise which will both highlight Bishop's continued activity in and academic emphasis of values in mathematics education, his influence in this aspect of mathematics education research, as well as other related research. This is thus not intended to be a review of research on values in mathematics education, and readers who are interested in such reviews may refer to Bishop, Seah and Chin (2003) and Seah, Atweh, Clarkson and Ellerton (2008).

The presentation and structure of the discussion in this chapter will reflect three different but related contexts, that of theorising values, "measuring" values, and applying what we know about values in school mathematics teaching and learning situations. The first section will examine how values in mathematics education have been theorised in recent years, looking at how regarding values as socio-cognitive in nature may help overcome problems associated with their being seen as affective variables. This section will also look at how the relationship between values and beliefs has been teased out in research, how Bishop's (1996) categorisation of values has evolved over the years, and revisit Bishop's (1999) observation that mathematics teaching/learning was often regarded by teachers as being value-free. The next section will discuss the relative difficulty with which values are identified and assessed in the school mathematics classroom. Last but not least, the last section will consider how the controversial nature of values teaching in mathematics lessons might be perceived, at a time when increasingly more nations and educational systems are investing in values education using a variety of different aspects of the school system.

Theorising Values in Mathematics Education

Values as an Affective Construct?

The construct of values has often been thought of as an affective variable. This might have been due to the categorisation of valuing as being the ultimate affective educational objective in the taxonomies presented by Krathwohl, Bloom, and Masia (1964). In this affective dimension of their three taxonomies, the characterisation of a learner's value complex is derived from successive internalisation of affective variables such as attitudes and beliefs. McLeod's (1992) chapter, however, distinguished amongst these variables in terms of the degree of cognition and affect involved, when he proposed that

> we can think of beliefs, attitudes, and emotions as representing increasing levels of affective involvement, decreasing levels of cognitive involvement, increasing levels of intensity of response, and decreasing levels of response stability. (pp. 578–579)

The progression of value indicators such as beliefs through seven criteria, as conceptualised by Raths, Harmin, and Simon (1987), is worth noting here as well. Five of these criteria, which involve the individual choosing and prizing particular values over others, certainly involve a good degree of cognitive processing. The positioning of values and of values research in the affective dimension of related research must have presented a theoretical barrier to any more meaningful understanding of how values are taught through school mathematics, and of how such teaching can be utilised also to enhance the effectiveness of mathematics pedagogy.

The inculcation, development, consideration and enactment of values within an individual's value system appear also to involve not just a cognitive or an affective dimension. Various research studies into the roles played by values in mathematics education have highlighted the extent to which these values can be – and are – subjected to sociocultural factors and influences as well.

One such source relates to the roles played by the teacher in a child's education. Reasonably, teachers who teach (mathematics) in foreign cultures (such as expatriate teachers and teachers on exchange programs) would be expected to experience the impact of new cultural values on their pedagogical decisions and actions. However, immigrant teachers of mathematics (in Australia) have also reported the experiencing of ongoing clashes between the values of their own socio-ethnic cultures and those of the host country (that is, Australia) in which they practise (Seah, 2004). These affected the professional lives of teachers who might have settled down physically in the host country, but who might nevertheless experience emotional dissonance when confronted with issues relating to how school mathematics can be effectively taught in their new homeland, and to a lesser but not insignificant extent, what or whose mathematics was to be shared with their students in class. Amongst "local teachers" too, we can find evidences of teachers' values underlying what and how they teach. Leu and Wu's (2002) study, for example, highlighted the extent to which a teacher's societal value (Confucianism) and another's religious value (Buddhism) play out in the respective teachers' pedagogical repertoire. Differences and similarities between the practices of the two Taiwanese teachers were explained through a comparison of Confucianism and Buddhism. These values of the two teacher participants were also contrasted with the Western culture and Christianity respectively.

It is, then, problematic to regard values and valuing as exclusively either a cognitive or an affective phenomenon. On the one hand, it is not just a function of mental processing, nor one which is solely governed by one's heartstrings. The sociocultural environment within which one finds oneself has been shown to impact on what and how values are internalised and acted upon. On the other hand, that valuing is a conscious process of choosing and deciding (and thus a cognitive process) is a view which needs to be balanced against the observation too that what we value can sometimes be stimulated by our inner sense of feeling and reckoning (an affective process). While McLeod's (1992) view quoted above implies that valuing exists along a cognition-affect continuum, there appears to be a need to posit this key process of human perception and learning within a socioculturally-rich space. The following excerpt illustrates this point forcefully:

> In the future, I want prospective mathematics teachers to read "The Victory Arc" (Froelich, 1991). In this article, it is demonstrated how to use graphing calculators to do mathematical modelling with no regard for ethical issues such as the morality of war. As an example of an applied problem-solving context devoid of consideration of larger societal issues, this article

> provides the impetus for critical reflection of particular values that mathematics implicitly defends and advocates. (Kitchen, 2005, p. 53)

In such a context, I would like to propose the possibility of considering the process of valuing as being a socio-cognitive one, which takes into account the factors discussed above, but without ignoring the affective dimension through its sociocultural lens. Perhaps more importantly, this will provide the valuing process with an identity and a sense of "academic belonging" in the mathematics education research community, something which seems to have eluded it since its conception by Bishop (1988). Being able to categorise valuing as socio-cognitive would hopefully provide it with a space from which further understandings and pedagogical activities might be developed.

Values and Beliefs

It is evident from the discussion above that values and beliefs are different in significant ways. Given that these two terms can be used interchangeably in daily usage, there have been several attempts at teasing out the relationship between values and beliefs. For example, Krathwohl et al.'s (1964) taxonomy regards values as evolving from beliefs. Bishop (2001) shared the same view when he made the observation that cultural influences on values are mediated by beliefs. Clarkson, Bishop, FitzSimons and Seah (2000) also defined values as beliefs in action, thereby locating values within the larger arena of beliefs.

Seah et al.'s (2008) assertion that "if a belief represents what one considers to be true, then a value represents the emphasis and importance one accords to related belief(s)" reflected an attempt at not just explicating how values and beliefs are related, but also how each appears individually. This differentiation of how beliefs and values might be identified developed from an earlier observation that beliefs are often expressed in contextualised forms, whereas the expression of what one values is often a de-contextualised one (see, for example, Seah, 2007). At the same time, these representations lend themselves well to relating how beliefs might evolve into values for any individual: The belief in what is true or valid may well be sufficiently validated over time, such that one grows to internalise and value this belief as being something of importance and significance. While a belief was necessarily contextualised in the various situations in which it is seen to be true, it is reasonable to regard the internalisation of the same belief and its subsequent metamorphosis into a value as being a process that strips away the different contexts within which the belief concerned is situated, so that what lies at the heart and core – the value – gets identified and embraced.

What would the implication for research be in the light of this clarification of the relationship between values and beliefs? Beliefs have been found in qualitative studies to influence teacher practice, and this has been validated in large-scale quantitative investigations as well. For example, Wilkins (2008) collected data from 481 primary school teachers, which led her to conclude that "comparing knowledge, attitudes, and beliefs, teachers' belief in the effectiveness of inquiry-based methods was found to have the strongest overall effect on teachers' use of inquiry-based instructional practices" (p. 156). It appears, then, that if values represent the deeply-held beliefs and they represent the core of what one considers to be important in one's life, then it might indeed be more effective and more productive to influence change (in ways of learning and of teaching) by facilitating value inculcation and development amongst students and teachers through their existing beliefs.

Value Categories

The discussion thus far does make values appear to be a relatively exclusive group of qualities within any individual's psyche. Yet, there are sufficient values for these to be perceived as being different and distinct from one another. For example, Bishop (1999) reiterated his earlier conception (Bishop, 1988) regarding the different groups into which values in the mathematics classroom in the Western world might be categorised. The general educational, mathematical and mathematics educational values reflect the "three principal sources of values in the mathematics classroom; society, mathematics, and mathematics education" (p. 3) respectively. It is worth noting, however, that these sources are by no means discrete and independent of one another. While Bishop's classification allows for the teasing out of the ideological, individual and social aspects of cultural phenomena to emphasise the breadth of value sources, it also provides space for particular values to be perceived as sharing multiple characteristics. As Seah (2004) noted,

> general educational, mathematical and specifically mathematics educational values do not exist mutually exclusive of one another. Some values fit into two or all three of the categories. For example, progress and its associated value of creativity can be as much a mathematical and mathematics educational value as a general educational value (depending on the socio-cultural context in which this is understood). (p. 45)

It must be noted that when Bishop (1996) conceptualised these three categories of values, he regarded them as values in mathematics classrooms, that is, the convictions and principles that are transacted in the teacher-student and student-student interactions taking place in mathematics classrooms. When we consider the teacher's work beyond the classroom, however, different kinds of values can begin to confront the teacher. Teacher planning and practice, for example, can be confronted by values representing the interest and emphasis of organisations like the school or education authorities. The Canadian immigrant teacher teaching in Australia in Seah's (2004) study reported that her pedagogical practice was constrained by a perceived relative lack of organisational valuing of *professional support*. For her, such support might include opportunities for teacher upskilling and availability of meaningful teacher support material. Clearly, an interpretation of organisational initiatives and policies can be enhanced by an explicit consideration of the underlying organisational values.

At the same time, any observed action or decision may not only reflect the valuing of more than one quality, but such qualities may also be grounded in multiple categories. Bills and Husbands (2005) related an episode in their observation of a secondary school mathematics class, in which a teacher's concern for the way her students responded to mistakes reflected a general societal value, general educational value, and a mathematical value. These different kinds of values were regarded as interacting amongst themselves in rich, complex and embedded ways.

Mathematics Education as a Value-Free Activity

Bishop (1999) highlighted the fact that the teaching and learning of school mathematics have been regarded by many, including teachers, as being value-free. "Most mathematics teachers would not even consider that they are teaching any values when they teach mathematics. Changing that perception may prove to be one of the biggest hurdles to be overcome" (Bishop, 1999, pp 1–2). That this was very much true in classrooms then was supported by research findings such as those of the *Values and Mathematics Project* (VAMP) (see Clarkson et al., 2000).

This is a significant situation in the broader context of school education, especially if many teachers recognise their professional roles and responsibility not just as teachers, but also as models of desirable values for their respective students. It raises questions relating to our conception of the nature of mathematics, and of school mathematics. If school mathematics is perceived as a body of cold hard facts and relationships, then it will be taught and learnt as a school subject that describes the world in absolute and pan-cultural terms. Rounding off numbers and decimals, for example, is often taught in ways that reflect and portray the message that there is one right way of accomplishing such a process irrespective of culture. Rarely do we come across a lesson in which the need for rounding off being dependent on certain cultures is discussed. Similarly, few teachers highlight the fact that the rounding-off "rules" do not apply in all cultures; for example, in the professional culture of fibre-optic engineering, (decimal) numbers ending with the digit 5 are not automatically rounded up, given that the overall bias of rounding up is significant in the context of precise measurements in that industry (see Fielding, 2003). In yet another professional or workplace culture, one which is represented at department store and supermarket cash registers, the smallest coin in circulation (in Australia, for example) or plain practical considerations (in Malaysia, for example) are just two of several factors underlying political decisions to round off cash transactions to the nearest five cents (instead of ten cents). In these contexts, then, how do we want our students to regard school mathematics, and the extent to which it models the world around us?

The observation that teachers and other stakeholders often regard school mathematics as value-free does point to a need for the nature of mathematics, and of school mathematics in particular, to be better understood in the society and in the education institutions. It is instructive that primary school teachers, amongst others, need to better appreciate and share with their students the different ways in which mathematical operations are conducted in different cultures. The exploration with students of the different ways in which multiple-digit multiplication is carried out (e.g. lattice method, Russian peasant algorithm, and the teachings inherent in vedic mathematics) will not only portray the valuing of *diversity* and *creativity*, say, but such pedagogical actions acknowledge the cultural knowledge and skills which ethnic minority students might bring with them to class. In addition, the introduction of alternative algorithms and ways of working mathematically can also be a source of challenge for the more able students to find the linkages between these alternative methods with those that are officially sanctioned and taught, thereby strengthening and deepening students' understanding of associated mathematical ideas in the process. At the same time, students are also given the opportunity to appreciate and understand how different mathematical practices are related to different ways in which different cultures perceive the environment around them.

There is evidence that the intended curricula in many cultures have been actively promoting the culture- and value-ladenness of (school) mathematics, and it is also probable that the curriculum developers and writers concerned are aware of their actions (see, for example, Bills & Husbands, 2005). An appreciation of this active promotion would be a crucial first step towards the facilitation of mathematics lessons in classes where students are given the opportunities more explicitly to explore and learn the values inherent in similarities and differences in the mathematical activities of different cultures. The curriculum statement for school mathematics in the state of Victoria in Australia (the Victorian Essential Learning Standards – Mathematics), for example, began with a statement that

> mathematics is a human endeavour that has developed by practice and theory from the dawn of civilisation to the present day. Many societies and cultures have contributed to the growth of mathematics, often in times of scientific, technological, artistic and philosophical change and development. Complementary to this broad perspective of mathematics are the various mathematical practices that take place day to day in communities around the world. (Victorian Curriculum and Assessment Authority, 2005, p. 4)

Encapsulated within this paragraph alone is a range of values related to mathematics and mathematical practices (e.g. *application, diversity, progress*) which could potentially be utilised by teachers in their lesson planning and execution. Perhaps, indeed, the curriculum statement in one's educational system might be a culturally-appropriate way for teachers to become more aware and conscious of the values they teach or portray.

However, within the Western nations with democratic traditions at least, it may be hard to envisage the intended curriculum going beyond relating mathematical ideas to activities undertaken in human civilisations. The previous quote is such an example, and values relating to mathematics and mathematical practices are often implied. Any more explicit reference to the teaching of values runs the risk of mathematics education being seen as hostage to acts of indoctrination. It is worth noting, however, that this is despite the fact that in many of these cultures, what each deems as desirable or important is often manifested in its own educational framework anyway (Gudmundsdottir, 1990; Kohlberg, 1981; Neuman, 1997). "Hence, it is not a question of whether education should deal with values. Education is about values inculcation and thus education cannot escape from dealing with value." (Atweh & Seah, 2008). At the level of classroom teachers, Bishop (1999) had noted that "we are also teaching students through mathematics. They are learning values through how they are being taught" (p. 4). Indeed, teachers

> cannot withdraw from showing the values that are important to them. In the cultural policy of the government and the school, teachers are even supposed to stimulate the development of specific values. But modern [western] society expects also more and more that young people make choices of their own accord and that they assume responsibility for these choices, also with regard to values. (Veugelers & Kat, 2000, p. 11, addition in brackets is mine)

Despite these developments over the last few years that have focused on values in the mathematics classroom, it is not clear as yet, in the absence of relevant recent research, whether classroom teachers of mathematics in general have become more aware of the value-laden nature of mathematics. Amongst teachers who acknowledge the roles played by values in mathematics teaching and learning, considerations and appraisals of how these are, or ought to be, portrayed are also not straightforward. Values are culturally relative, and even for those values which are often perceived as culturally universal (such as *respect*), they can be operationalised in different ways across *and* within different cultures. Indeed, this is one aspect underlying the difficulty with which values might be "measured", both in research and for educational management.

"Measuring" Values in Mathematics Education

Measuring is involved with the quantifying of qualities. The nature of values makes such quantifying understandably subjective and arguable. What is valued by teachers may be "expressed in the content of their instruction and in the way they guide the learning process" (Veugelers, 2000, p. 40), but how might one's valuing of *creativity*, say, be measured in ways that are objective, valid and reliable? This is perhaps one reason why most, if not all, values in mathematics education have been investigated through qualitative research designs utilising ethnographic and case study approaches (e.g. Seah, 2004). On the other hand, this also makes it difficult for school leaders and classroom teachers to engage in the assessment of values teaching in (mathematics) lessons in quick, efficient ways. For educational leaders and teachers, the use of qualitative collection and analysis of data is both a luxury constrained by tight time demands, as well as an assessment technique which relatively few people have the confidence to harness. Hence it is rarely seen as an option by teachers in classrooms.

Furthermore, the deployment of any qualitative methodology to identify values validly calls for the use of multiple methods. As discussed above, any observed action may reflect one or more values, not all of which are necessarily held by or subscribed to by an individual. The identification of what aspects are valued by the individual comes from a process of triangulating the data obtained from multiple sources. For examples, Clarkson et al.'s (2000) discussion of the difficulties involved in researching values, Seah's (2004) work with immigrant teachers of mathematics, as well as Galligan's (2005) investigation of teacher values, are examples of studies which involved the use of questionnaires, lesson observations, interviews *and* document analyses. A classroom teacher who wishes to find out what a particular student values will thus need to do more than, say, observe what the student does or says. The teacher may need to clarify with the student to confirm, or he/she may cross-check with further observations of the student's behaviour. These are indeed by no means easy or convenient ways in which values might be "measured".

One possibility that might be worth exploring is to "measure" values through assessing one's beliefs. This approach builds on the observation that beliefs are already mapped reliably using established instruments, and on the close relationship between one's beliefs and one's value system (as discussed above). In fact, they may be regarded as influencing each other's development. Krathwohl et al.'s (1964) taxonomy, for example, conceptualises values as deriving from beliefs. The extent to which values are internalised within an individual, however, makes it reasonable too to consider that what one regards as important (associated with values) affects what one considers to be true (associated with beliefs). For example, a teacher who values *technology* as being an important tool to support (mathematics) learning and teaching can be imagined to subscribe to beliefs such as, *the use of interactive whiteboards helps children to become more involved in working mathematically in the classroom*, as well as *children calculator use frees up time for children to engage in higher-order thinking activities*.

A point to note, however, is that while beliefs can imply the types of values an individual subscribes to, they do not imply what an individual does not value. The function of competing values (Seah, 2004) would mean that depending on the context and its associated constraints and affordances, a value can be overridden by another one held by the same individual, which is subsequently operationalised either as an action/decision, or as a belief. Nevertheless, there appears to be potential for one's beliefs to be analysed as a group, such that the underlying values subscribed to might be inferred, and cross-checked if necessary.

Applying Research Knowledge

The relative difficulty with which values can be measured or assessed in the typical classroom is an example of the barriers of more widespread explicit use of and reference to values in the mathematics classroom.

Regardless of the academic knowledge and understandings we might have accumulated, any application of these in school practice (through intervention or otherwise) is less than straightforward. As alluded to in the last section, there is certainly the debate about the legitimacy for the teaching of values in school education, especially in state/government schools (as opposed to religious based schools), which are often expected to deliver a secular curriculum to children. Even in educational contexts where values teaching is expected or recognised, there are also the complex issues relating to the portrayal or transmission of values which are not intrinsic to particular subjects, that is, general societal values which are not the subject values – in this case mathematical values (Halstead & Taylor, 2000). Thus, it can be controversial, say, for the valuing of *(racial) harmony* to be promoted in mathematics lessons, as values such as this do not make up what mathematics is, and teachers or education systems might not be expected to exploit students' passion or learning of the subject to inject more general values.

However, particularly in recent years, there appears to be a greater and more evident push at the governmental level in many countries for the teaching of "desirable" values to be effected through school education (see examples below). This development often runs complementarily to the promotion in mathematics curriculum documents of the cultural aspect of mathematics teaching and learning (as discussed earlier in this chapter). Although it is generally acknowledged in many of these official campaigns that values are portrayed and transmitted through school education, the purpose of the initiatives appears to be the encouragement for nationally "desirable" values to be highlighted. Examples of such governmental initiatives include the *National framework for values education in Australian schools* (Department of Education, Science and Training, 2005) in Australia, the *Manifesto on values, education and democracy in the National Education program* (South Africa Department of Education, 2001) in South Africa, the articulation of a set of *Shared Values* (Singapore Government, 1991) in Singapore, and the *Budi Bahasa dan Nilai-Nilai Murni* (courtesy and noble values) in Malaysia (Tan, 1997). It is likely that the increased explicitness of these policies represents some countries' responses to perceived threats to the preservation of national cultures and ways of living as a result of globalisation of activities at all levels of modern-day living. In such contexts, there are actually more opportunities for teachers of mathematics at all school levels to consider how and what values may be taught through their pedagogical repertoire, as well as how the inculcation of relevant values might foster more positive student dispositions towards – and greater student understanding of – the subject.

Closing Remarks

Such attempts by teachers to optimise the mathematics learning experiences through values teaching, and to optimise values inculcation by students through mathematics learning, are best supported through a sustained effort in research and associated development. This is the intersection in need of research which Bishop (1999) identified. Over the years, mathematics educational research investigating the roles played by the interaction of values and mathematics in school education has certainly increasingly been taken up by more researchers over a greater geographical area, covering America (e.g. Dahl, 2005), Asia (e.g. Chin, Leu, & Lin, 2001),

Australia (e.g. Galligan, 2005; Seah, 2007), and Europe (e.g. Bills & Husbands, 2005; Hannula, 2002). Bishop (1999) might have noted that "from a research perspective the International Handbook on Mathematics Education . . . is revealing. It has no specific chapter on values" (p. 3); and he and other interested researchers might take heart in the knowledge that within a decade of that observation, not only was a chapter (Bishop et al., 2003) dedicated to values included in that handbook's second edition, there have also been significant book chapters (e.g. Bishop & Seah, in press) published in this area. It is worth noting too the opportunities for such research to be conducted together with other aspects of school education, such as in early childhood (e.g. Court & Rosental, 2006), with science (e.g. Bishop, Clarke, Corrigan & Gunstone, 2005) and history (e.g. Bills & Husbands, 2005). Indeed, as mathematics education research transforms itself into a form that a related and ongoing regional research project in Australia/East Asia labels as the third wave (following on from earlier waves of cognitive and affective foci to a socio-cognitive focus; see notes below), there is every reason for all stakeholders in mathematics education and its research to expect a greater recognition of the roles that values and mathematics pedagogy play in complementing student learning of each of these areas in the near future.

Given the relationship between values and beliefs, and given current knowledge of the strong effect that beliefs exert on teacher practice (Leder, Pehkonen, & Torner, 2002; Wilkins, 2008), there continues to be a need for the intersection of mathematics teaching and values education to be understood more deeply and researched further. Despite the academic progress made in the research area of values in mathematics education in the years since Bishop (1999) identified this need, and with the confidence that the researching of the socio-cognitive variable of values in mathematics teaching and learning will continue to expand, there appears to be a need for even more far-sighted institutions to provide the necessary recognition and support to make the breakthrough in improving the quality of mathematics pedagogy in schools through the socio-cognitive construct of values today. Negotiating this intersection does not call for any flyover, tunnel or bypass to be constructed to facilitate the respective journeys that mathematics teaching and values education undertake separately. Instead, it requires all stakeholders on these journeys to pause, enrich and rejuvenate one another at this intersection before they continue their respective quests to learn/teach associated knowledge, skills and dispositions in mathematics and values in more effective and empowering ways.

Notes

The third wave: A regional study of values in effective numeracy education is a research project being piloted in 2008–2009 across six regions, that is, in Australia (Tamsin Meaney & Wee Tiong Seah), Hong Kong SAR (Ngai Ying Wong), Malaysia (Chap Sam Lim), Singapore (Siew Yin Ho), Taiwan (Chien Chin) and Vietnam (Tran Vui). This project aims to map, compare and contrast the convictions and preferences co-valued by teachers and students in effective numeracy lessons across the six regions, and between indigenous and immigrant contexts within some of these.

References

Atweh, B., & Seah, W. T. (2008, February 1). *Theorizing values and their study in mathematics education*. Paper presented at the Australian Association for Research in Education Conference, Fremantle, Australia. Retrieved April 5, 2008, from http://www.aare.edu.au/07pap/atw07578.pdf.

Bills, L., & Husbands, C. (2005). Values education in the mathematics classroom: subject values, educational values and one teacher's articulation of her practice. *Cambridge Journal of Education, 35*(1), 7–18.

Bishop, A. J. (1988). *Mathematical enculturation: a cultural perspective on mathematics education*. Dordrecht, The Netherlands: Kluwer Academic Publishers.

Bishop, A. J. (1996, June 3–7). *How should mathematics teaching in modern societies relate to cultural values – some preliminary questions*. Paper presented at the Seventh Southeast Asian Conference on Mathematics Education, Hanoi, Vietnam.

Bishop, A. J. (1999). Mathematics teaching and values education: an intersection in need of research. *Zentralblatt fuer Didaktik der Mathematik, 31*(1), 1–4.

Bishop, A. J. (2001). What values do you teach when you teach mathematics? *Teaching Children Mathematics, 7*(6), 346–349.

Bishop, A., Clarke, B., Corrigan, D., & Gunstone, D. (2005). Teachers' preferences and practices regarding values in teaching mathematics and science. In P. Clarkson, A. Downton, D. Gronn, M. Horne, A. McDonough, R. Pierce, & A. Roche (Eds.), *Building connections: research, theory and practice* (Vol. 1, pp. 153–160). Sydney, Australia: Mathematics Education Research Group of Australasia.

Bishop, A. J., & Seah, W. T. (in press). Educating values through mathematics teaching: possibilities and challenges. In M. H. Chau & T. Kerry (Eds.), *International perspectives on education* (pp. 118–138). London, UK: Continuum.

Bishop, A. J., Seah, W. T., & Chin, C. (2003). Values in mathematics teaching: the hidden persuaders? In A. J. Bishop, K. Clements, C. Keitel, J. Kilpatrick & F. Leong (Eds.), *International handbook of mathematics education* (2nd ed., pp. 715–763). Dordrecht, The Netherlands: Kluwer Academic Publishers.

Chin, C., Leu, Y.-C., & Lin, F.-L. (2001). Pedagogical values, mathematics teaching, and teacher education: case studies of two experienced teachers. In F.-L. Lin & T. J. Cooney (Eds.), *Making sense of mathematics teacher education* (pp. 247–269). Dordrecht, The Netherlands: Kluwer Academic Publishers.

Clarkson, P., Bishop, A. J., FitzSimons, G. E., & Seah, W. T. (2000). Challenges and constraints in researching values. In J. Bana & A. Chapman (Eds.), *Mathematics education beyond 2000: Proceedings of the twenty-third annual conference of the Mathematics Education Research Group of Australasia Incorporated held at Fremantle, Western Australia, 5–9 July, 2000.* (Vol. 1, pp. 188–195). Perth, Australia: Mathematics Education Research Group of Australasia Incorporated.

Court, D., & Rosental, E. (2006). Values embodied in children's literature used in early childhood education in Israeli state schools. *Early Childhood Education Journal, 34*(6), 407–414.

Dahl, B. (2005). A comparison of the Danish and the Virginia secondary teacher education system: their values and emphasis on mathematics content knowledge. *The Montana Mathematics Enthusiast, 2*(2), 93–106.

Department of Education, Science and Training (2005). *National framework for values education in Australian schools*. Canberra, Australia: Department of Education Science and Training.

Fielding, P. (2003). Are we teaching appropriate rounding rules? In B. Clarke, A. Bishop, R. Cameron, H. Forgasz, & W. T. Seah (Eds.), *Making mathematicians* (pp. 395–401). Victoria, Australia: The Mathematical Association of Victoria.

Froelich, G. (1991). The victory arc. *Consortium: the newsletter for the consortium and its applications, 38*, 2–5.

Galligan, L. (2005). Conflicts in offshore learning environments of a university preparatory mathematics course. In H. Chick & J. L. Vincent (Eds.), *Proceedings of the 29th conference of the International Group for the Psychology of Mathematics Education* (Vol. 3, pp. 25–32). Melbourne: PME.

Gudmundsdottir, S. (1990). Values in pedagogical content knowledge. *Journal of Teacher Education, 41*(3), 44–52.

Halstead, J. M., & Taylor, M. J. (2000). *The development of values, attitudes and personal qualities: a review of recent research*. Slough, UK: National Foundation for Educational Research.

Hannula, M. S. (2002). Attitude towards mathematics: emotions, expectations and values. *Educational Studies in Mathematics, 49*, 25–46.

Kitchen, R. S. (2005). Making equity and multiculturalism explicit to transform mathematics education. In A. J. Rodríguez & R. S. Kitchen (Eds.), *Preparing mathematics and science teachers for diverse classrooms: promising strategies for transformative pedagogy* (pp. 33–60). New York, USA: Lawrence Erlbaum Associates.

Kohlberg, L. (1981). *Philosophy of moral development : moral stages and the idea of justice* (1st ed.). San Francisco, USA: Harper & Row.

Krathwohl, D. R., Bloom, B. S., & Masia, B. B. (1964). *Taxonomy of educational objectives: the classification of educational goals (Handbook II: Affective domain)*. New York USA: David McKay.

Leder, G. C., Pehkonen, E., & Torner, G. (Eds.) (2002). *Beliefs: a hidden variable in mathematics education*. Dordrecht, The Netherlands: Kluwer Academic Publishers.

Leu, Y.-C., & Wu, C.-J. (2002). The origins of pupils' awareness of teachers' mathematics pedagogical values: Confucianism and Buddhism-driven. In *Pre-conference proceedings of the ICMI Comparative Study Conference 2002 Hong Kong* (pp. 49–55). Hong Kong: The University of Hong Kong.

McLeod, D. B. (1992). Research on affect in mathematics education: a reconceptualisation. In D. A. Grouws (Ed.), *Handbook of research on mathematics teaching and learning*. (pp. 575–596). Reston, VA, USA: National Council of Teachers of Mathematics.

Neuman, W. L. (1997). *Social research methods: qualitative and quantitative approaches* (3rd ed.). Boston, MA, USA: Allyn & Bacon.

Raths, L. E., Harmin, M., & Simon, S. B. (1987). Selections from 'values and teaching'. In J. P.F. Carbone (Ed.), *Value theory and education* (pp. 198–214). Malabar, FL, USA: Robert E. Krieger.

Seah, W. T. (2004). *The negotiation of perceived value differences by immigrant teachers of mathematics in Australia*. Unpublished PhD dissertation, Monash University, Victoria, Australia.

Seah, W. T. (2007). Qualities co-valued in effective mathematics lessons in Australia: preliminary findings. In J. H. Woo, H. C. Lew, K. S. Park, & D. Y. Seo (Eds.), *Proceedings of the 31st conference of the International Group for the Psychology of Mathematics Education* (Vol. 4, pp. 161–168). Seoul, South Korea: International Group for the Psychology of Mathematics Education.

Seah, W. T., Atweh, B., Clarkson, P., & Ellerton, N. (2008). Socio-cultural factors in mathematics teaching and learning. In H. Forgasz, T. Barkatsas, A. Bishop, B. Clarke, S. Keast, W. T. Seah, P. Sullivan, & S. Willis (Eds.), *Research in Mathematics Education in Australasia 2004–2007* (pp. 223–253). Rotterdam, The Netherlands: Sense Publishers.

Singapore Government. (1991). *Shared values* (White paper No. Cmd 1 of 1991). Singapore: Government of Singapore.

South Africa Department of Education. (2001). *Manifesto on values, education and democracy*. Pretoria, South Africa: South Africa Department of Education.

Tan, S. K. (1997). Moral values and science teaching: a Malaysian school curriculum initiative. *Science and Education, 6*, 555–572.

Veugelers, W. (2000). Different ways of teaching values. *Educational Review, 52*(1), 37–46.
Veugelers, W., & Kat, E. D. (2000, April 24–28). *The teacher as a moral educator in secondary education: the student perspective*. Paper presented at the 81st Annual Meeting of the American Educational Research Association, New Orleans, LA.
Victorian Curriculum and Assessment Authority. (2005). *Discipline-based learning strand: mathematics*. East Melbourne, Victoria, Australia: Victorian Curriculum and Assessment Authority.
White, L. A. (1959). *The evolution of culture*. New York: McGraw-Hill.
Wilkins, J. L. M. (2008). The relationship among elementary teachers' content knowledge, attitudes, beliefs, and practices. *Journal of Mathematics Teacher Education, 11*, 139–164.

Index

A
Abas, S. J., 145
Abdeljaouad, M., 143, 144
Abreu, G. D., 20
Adam, S., 126, 127
Adams, B., 126
Adler, J., 129
Alangui, W., 125, 128
An, S., 141, 142
Anastasi, A., 75
Anderson, M., 93
Arfwedson, G., 194, 195
Ascher, M., 124, 125
Askew, M., 217, 219
Askwe, J. M., 137
Aspinwall, L., 91
Atweh, B., 240, 246

B
Baker, D., 219
Balacheff, N., 205, 211
Balchin, W. G., 79
Ball, D. L., 56, 57, 58, 60, 62, 63, 64
Ball, D., 57
Barakat, M. K., 71
Barton, B., 6, 107, 121, 123, 124, 126, 128, 129
Barwell, R., 122, 129
Bass, H., 56, 57, 64
Battey, D., 61
Battista M. T., 92, 198
Beaton, A. E., 138
Becker, J. P., 77
Begle, E. G., 196
Behr, M. J., 197
Bender, L., 112
Berggren, J. L., 143, 144
Berliner, D. C., 52, 53, 54
Berry, J. W., 75
Biagetti, S., 61
Bibby, T., 219
Biddle, B. J., 227
Biersack, A., 116, 117
Biggs, J. B., 137, 139
Bills, L., 244, 245, 249
Bishop, A., 4, 5, 6, 7, 9, 10, 13–25
Bishop, A. J., 27, 28, 30, 31, 37–63, 71, 76, 77, 83, 84, 85, 86, 87, 91, 92, 93
Blane, D. C., 207, 208
Bloom, B. S., 98, 233, 240
Bloor, D., 122
Boero, P., 207
Bonnevaux, B., 102
Booth, S., 144
Borko, H., 6, 16, 19, 27, 37, 44, 51, 52, 53, 54, 55, 57, 63, 64
Boston, M. A., 13, 19
Brimer, A., 139
Brinkmann, E. H., 77
Brophy, J., 62, 193, 198
Brown, F. R., 78
Brown, M., 217, 218, 220, 221, 222
Brown, S. I., 236
Brown, T., 122
Bruner, J. S., 13, 14, 15, 74
Bryant, P. E., 73
Burstein, L., 51
Buxton, L., 234

C
Calderhead, J., 53, 54
Carr, W., 196, 197, 198
Carry, L. R., 76
Carson, R., 124, 126
Chen, C., 137
Chin, C., 5, 230, 240, 248, 249
Christine, K., 6, 167, 187, 207
Chrostowski, S. J., 138
Cifarelli, V. V., 93

Clark, C. M., 44, 51
Clarke, B., 249
Clarke, D., 7
Clarke, E., 115
Clarkson, P. C., 229, 230, 239, 242, 244, 247
Clements, D. H., 198
Clements, K., 92
Clements, M. A., 7, 19, 20, 69, 78
Cockburn, A., 198, 199, 200, 201, 208
Cockcroft, W. H., 215, 216, 224, 232
Cohen, D. K., 62, 63, 64
Cole, M., 76, 102
Contreras, L. O., 123
Cooke, M., 123
Cooney, T. J., 193
Corrigan, D., 249
Court, D., 249
Crosswhite, F. J., 193
Crowe, M. J., 122

D
D'Ambrosio, U., 122, 123, 125, 168
Dahl, B., 248
Damerow, P., 168, 171, 172, 179, 192
Daubin, J. W., 122
Davis, B., 122
Davis, P., 178
Davis, R. B., 193
Dawson, J. L. M., 77
de Oliveira, C., 126
De Young, G., 143
Denvir, H., 219
Deregowski, J. B., 101, 110, 121
Desforges, C., 198, 199, 200, 201, 208, 220, 226
Domite, M., 123
Donaldson, M., 73, 115
Dowling, P., 122, 217
Downs, R. M., 8
Doyle, W., 198

E
Earl, L., 220, 221, 222, 223
Eisenhart, M., 51
Eiteljorg, E., 63
Ellerton, N., 125, 240
Ellerton, N. F., 198
Elliott, J., 216
Elstein, A. S., 38
Elwitz, U., 172
Englund, R. K., 171
Erickson, F., 58
Ernest, P., 236
Ethington, C. A., 234

F
Fasheh, M., 234
Fehr, H., 209
Fennema, E., 75, 76
Fielding, P., 244
Finley, G. E., 102
Fischbein, E., 79
FitzSimons, G. E., 242
Fou-Lai, L., 18, 230
Frandsen, A. N., 77
Franke, M. L., 61, 62
Frankenstein, M., 126
Frederick, L., 8, 135
Freudenthal, H., 8, 93, 103, 193, 209
Fullan, M., 222

G
Galligan, L., 129, 247, 249
Galton, F., 71
Garden, R. A., 137, 234
Gay, J., 76
Gerdes, P., 124, 125, 168
Gilead, S., 91
Glaymann, M., 209
Golafshani, N., 144
Gonzales, M., 102
Gonzales, P., 138
Good, T. L., 227
Gooya, Z., 143, 144
Gravemeijer, K., 86, 93
Greeno, J. G., 53
Griffin, P., 139
Griffin, R., 143
Grossman, P., 55
Grouws, D. A., 92, 193, 197
Gu, L., 144
Guay, R. B., 72, 73
Gudmundsdottir, S., 246
Gunstone, D., 249
Gutierrez., 87

H
Hadamard, J., 75
Halliday, M. A. K., 129
Halstead, J. M., 248
Han, S. Y., 63
Hanna, G., 144
Hannula, M. S., 249
Harmin, M., 233, 240
Harris, L. T., 75
Hart, K. M., 139
Hart, R. A., 73
Hashweh, M. Z., 55
Hedger, K., 75

Hersh, R., 129, 178
Hiebert, J., 140
Hill, H. C., 56, 57, 58, 64
Hirsch-Dubin, F., 126
Ho, D. Y. F., 132, 249
Horng, W.S., 230
Horsthemeke, K., 124, 126
Howson, A. G., 235
Howson, G., 136
Howson, G. A., 176
Huang, R., 140, 144
Huberman, M., 225
Husbands, C., 244, 245, 249

I
Inhelder, B., 73

J
Jablonka, E., 179
Jacobs, J., 61, 63
Jahnke, T., 182
Jeremy, K., 8, 9, 17, 19, 205
Johnson, D. C., 219
Johnson, M., 128
Jones, D., 193
Jones, J., 105, 116
Joseph, G. G., 122, 137, 144

K
Kagan, J., 102
Kazemi, E., 61, 62
Kearins, J., 101
Keitel, C., 7, 92, 125, 136, 139, 150, 167, 168, 172, 176, 178, 180, 181
Kello, M., 51
Kelly, G. A., 210
Kelly, M., 113, 116
Kemmis, S., 196, 197, 198
Kennedy, J. M., 89, 114
Kennedy, W. A., 72
Kent, D., 75
Kieran, C., 197
Kilpatrick, J., 176, 181, 189, 193
Kitchen, R. S., 242
Klein, R. E., 8, 102
Knijnik, G., 126
Kohlberg, L., 234, 246
Kotzmann, E., 178
Krathwohl, D. R., 233, 240, 242, 247
Krutetskii, V. A., 9, 72, 73, 75, 85, 87, 88, 89, 105
Kulm, G., 141

L
Laborde, C., 92
Lakoff, G., 128
Lancy, D. F., 75, 102, 115
Land, F. W., 98
Lapointe, A. E., 137
Latterall, C. M., 122
Lave, C., 102
Lean, G. A., 19, 20, 22, 72, 85, 87
Lean, G., 113
Leder, G. C., 236, 249
Lee, S., 137
Lee, S. Y., 137
Lef'evre, W., 172
Lehrer, R., 86, 231
Leinhardt, G., 53
Leithwood, K., 222
Lerman, S., 181
Lesh, R., 73, 74
Lester, F. K. Jr., 93
Leu, Y.-C., 241, 248
Leung, F., 107, 135, 139, 140, 141, 142, 144
Leung, F. K. S., 8
Levin, B., 222
Lin, F. L., 18, 139, 230, 248
Lipka, J., 126
Livingston, C., 53, 54
Lortie, D. C., 162
Loveless, T., 125
Lummis, M., 137

M
Ma, L., 141, 142
McDaniel, E. D., 72
McGee, M. G., 72, 75
McIntyre, D., 98
McIntyre, D. I., 218
McLeod, D. B., 234, 240, 241
Maclure, M., 216
Magne, O., 75
Makdisi, G., 144
Mariotti, M. A., 87
Marriott, P., 77
Martin, J. L., 73
Martin, M. O., 138
Marton, F., 144
Masia, B. B., 233, 240
McTaggart, T., 196, 197
Mead, N. A., 137
Metzler, J., 78, 91
Meyerhöfer, W., 182
Michael, W. B., 72
Millett, A., 219

Mitchelmore, M. C., 76, 78
Moore, G. T., 73
Morales, J. F., 123
Muijs, D., 220
Mullis, I. V. S., 138
Munro, D., 101

N
Neisser, U., 79, 91
Neuman, W. L., 246
Nickson, M., 214
Niles, J. A., 51
Nissen, H., 171
Nolan, E., 102
Noss, R., 217

O
O'Halloran, K. L., 129
Owens, K., 90

P
Park, K., 139, 140, 142
Park, K. S., 205
Parker, T., 102
Parzysz, B., 91
Pehkonen, E., 249
Peirce, C. S., 91, 92
Peterson, P. L., 44, 51
Phelps, G., 56, 58
Philp, H., 19, 113, 116
Piaget, J., 73, 87, 97, 103
Pinxten, R., 122
Pittman, M. E., 63
Poisard, C., 123
Popkewitz, T. S., 197, 220
Powell, A. B., 126
Presmeg, N. C., 18, 20
Presmeg, N., 3, 9, 88, 89, 90, 91, 93
Pritchard, C., 86
Putnam, R., 55, 63, 64

Q
Quinton, M., 234

R
Radatz, H., 76, 77
Ramírez, J. F., 123
Ranucci, E. R., 78
Raths, L. E., 233, 234, 240
Ray, W., 128
Renn, J., 173
Renuka, V., 11, 18, 167, 187
Restivo, S., 122
Reston, V. A., 151, 152
Reynolds, D., 220
Rhodes, V., 219
Richard, S., 37, 38
Richardson, A., 91
Richert, A. E., 54
Roberts, S. A., 9, 16, 27, 37, 57
Robitaille, D. F., 137
Rofagha, N., 144
Rogoff, B., 102
Rosental, E., 249
Ross, A. S., 114
Rowan, B., 56
Rowlands, S., 124, 126
Ruthven, K., 9, 23, 187, 189, 213, 214, 218, 226

S
Sáenz-Ludlow, A., 93
Sarland, C., 216
Saunderson, A., 77
Schäfer, M., 124, 126
Schilling, S., 58
Schoenfeld, A. H., 198
Scribner, S., 102
Seago, N., 62
Seah, T.,
Seah, W. T., 10, 24, 25, 230, 239, 241, 242, 243, 246, 247, 248, 249
Senft, G., 122
Setati, M., 122, 129
Sharp, D., 102
Sharp, F., 126
Sharp, N., 126
Shaughnessy, J. M., 198
Shavelson, R. J., 10, 16, 37, 38, 39, 44, 45, 46, 47, 48, 49, 50, 51, 52, 53, 54, 64
Shaw. K. L., 91
Sheehan, P. W., 79
Shepard, R. N., 78, 91
Sherin, M. G., 62, 63
Sherman, J. A., 78
Shih, J., 61
Shimizu, Y., 7, 180
Shternberg, B., 91
Shulman, L. S., 38, 39, 43, 44, 46, 47, 49, 50, 54, 55, 56, 64, 97
Sierpinska, A., 176, 207, 211, 213
Simon, S. B., 99, 233
Skovsmose, O., 126, 128, 175, 178, 183, 232, 236
Smith, A., 71
Smith, D., 53
Smith, I. M., 98

Soberon, E., 122
Sosniak, L.A., 234
Spearman, C. E., 71
Spengler, O., 122
Sprafka, S. A., 38
Stanley, J. C., 75
Steffe, L. P., 90
Stern, P., 44, 45, 47, 49
Stevenson, H. W., 102, 137, 138
Stigler, J., 137
Stigler, J. W., 140
Stoll, L., 221, 223
Strathern, A., 109
Sullivan, E. V., 74
Sun, L. K., 137
Szeminska, A., 73

T
Taher, M., 143
Tall, D., 128
Tan, S. K., 248
Taylor, M. J., 248
Thames, M. H., 56, 58
Thompson, A., 234
Thompson, P. W., 90
Thurstone, L. L., 71, 73, 85
Tomlinson, P., 234
Torner, G., 249
Trabasso, T., 73
Tsatsaroni, A., 181

U
Umland, K., 129

V
Van Dooren, I., 122
Van Dormolen, J., 230
Van Hiele, P. M., 74, 87
Van Oers, B., 86
Vandett, N., 51
Varelas, M., 234
Verschaffel, L., 86
Veugelers, W., 246
Vithal, R., 11, 18, 125, 126, 150, 167, 183, 187
Vladimirskii, G. A., 77

W
Waghid, Y., 184
Wales, R. J., 115
Wanderer, F., 126
Washington, D. C., 151, 205
Watson, N., 139, 222
Wattanawaha, N., 73, 99
Webb, L. F., 76
Webb, N. M., 11, 51
Werdelin, I., 71, 72
Werner, H., 73, 74, 86, 87
Wheatley, G. H., 75, 91
White, L. A., 235, 239, 247
Whitfield, R., 15, 16, 19, 21, 22, 27, 28
Wilder, R. L., 122, 192
Wiliam, D., 217
Wilkinson, A., 102
Wilson, B., 137
Wilson, B. J., 235
Wilson, S. M., 54, 55
Woodrow, D., 178, 181
Wrigley, J., 71
Wu, C.-J., 241
Wu, Z., 141

X
Xu, G., 181

Y
Yerushalmy, M., 91
Young, C. D., 77

Z
Zaslavsky, C., 126, 145
Zellweger, Z., 93
Zimmer, J., 172

FSC
www.fsc.org
MIX
Papier | Fördert
gute Waldnutzung
FSC® C083411

Kolibri 360 GmbH
Ferdinand-Jühlke-Straße 7
99095 Erfurt, Deutschland
produktsicherheit@kolibri360.de